Monoclonal Antibody-Directed Therapy

Monoclonal Antibody-Directed Therapy

Editors

Veysel Kayser
Amita Datta-Mannan

MDPI • Basel • Beijing • Wuhan • Barcelona • Belgrade • Manchester • Tokyo • Cluj • Tianjin

Editors
Veysel Kayser
Sydney School of Pharmacy
The University of Sydney
Sydney
Australia

Amita Datta-Mannan
Exploratory Medicine &
Pharmacology
Eli Lilly and Company
Indianapolis
United States

Editorial Office
MDPI
St. Alban-Anlage 66
4052 Basel, Switzerland

This is a reprint of articles from the Special Issue published online in the open access journal *Antibodies* (ISSN 2073-4468) (available at: www.mdpi.com/journal/antibodies/special_issues/mAb_therapy).

For citation purposes, cite each article independently as indicated on the article page online and as indicated below:

LastName, A.A.; LastName, B.B.; LastName, C.C. Article Title. *Journal Name* **Year**, *Volume Number*, Page Range.

ISBN 978-3-0365-2873-1 (Hbk)
ISBN 978-3-0365-2872-4 (PDF)

Contents

About the Editors

Veysel Kayser

Veysel Kayser is an Associate Professor at The University of Sydney (USyd). He completed his B.Sc. degree in Chemistry at Hacettepe University (Turkey) and his Ph.D. in Physical Chemistry from the University of Leeds (UK) in 2004. Subsequently, he undertook post-doctoral fellowships at the Max-Planck Institute of Biochemistry (Germany) and in the Department of Chemical Engineering at MIT (US), and then he worked at MIT for several years as a senior staff scientist. In mid-2013, he took up his current position at USyd. For periods, he was also the Associate Dean for Research and the HDR coordinator at the Faculty of Pharmacy and served in the Multidisciplinary Advisory Board of the Marie Bashir Institute for Infectious Diseases and Biosecurity at USyd (now Sydney Institute for Infectious Diseases).

His research interests are in biologics and vaccines (influenza, rabies, COVID-19 etc.), development of biosimilars and biobetters–mainly therapeutic monoclonal antibodies (mAbs), antibody-drug-conjugates, nanoparticle-antibody complexes, protein engineering, novel formulations of biologics and vaccines, and investigating the mechanisms of protein unfolding and aggregation.

He has obtained five patents, and several patent applications are pending, published over fifty peer-reviewed research publications, has given over 100 talks on biologics and vaccines, and also serves as an editor or member of the editorial board of various journals. He has consulted for and received research funding from industry. He has received numerous awards and fellowships.

Amita Datta-Mannan

Amita Datta-Mannan is a Research Fellow and Clinical Pharmacologist in the Department of Exploratory Medicine and Pharmacology at Eli Lilly and Company. She earned her Bachelor of Science at McMaster University, Ontario, Canada, in 1998 and her Ph.D. in Protein and Physical Chemistry from Brown University and Indiana University in 2003. Dr. Datta-Mannan has over 17 years of experience as a pharmaceutical scientist leading cross-functional drug discovery and development teams in various technical and corporate leadership roles. She is passionate about enabling the progression of novel therapeutics and medicines to improve the quality of life. Along these lines, Dr. Datta-Mannan has contributed to discovery and development of multiple medicinal modalities including monoclonal antibodies, bispecific antibodies, antibody drug conjugates, peptides and fusion proteins for autoimmunity, oncology, diabetes, migrane and musculoskeletal disorders. Dr. Datta-Mannan has key platform expertise in the area of mechanisms influencing the disposition, biodistribution and metabolism of biologics following parenteral and oral delivery. She also developed the first surrogate chemokine receptor structures to dissect the structure-function relationship of receptor-ligand interactions for developing therapies directed at autoimmune disorders. Dr. Datta-Mannan has contributed to multiple patents, published over 50 peer reviewed manuscripts and abstracts, including invited scientific reviews and pharmaceutical industry whitepaper guidance documents for biologics, presented/chaired over 30 invited oral presentations and serves on the editorial/review board of several journals.

Preface to "Monoclonal Antibody-Directed Therapy"

The unparalleled specificity and high efficacy of monoclonal antibodies (mAbs) make them desirable modalities both as biological medicines and diagnosis tools. They are amongst the top-selling drugs globally and their market size continues to grow annually. There are over 100 mAb products on the market currently, but considering the thousands that are in clinical trials and the new advancements in this field, such as the development of biosimilars, their market share is expected to increase substantially in the near future. Developing an antibody therapeutic, however, is an onerous journey, as many degradation pathways can prevent this process occurring successfully. These roadblocks generally present themselves at every stage of drug development.

With the advancement of the field over recent years, different types of complex mAb formats have been developed, including full-size mAbs, antibody fragments, antibody-drug-conjugates (ADC) and bispecifics, but full-size mAbs by far still dominate the market. In addition, new approaches such as PEGylation or the hyperglycosylation of constant domains of mAbs and other forms of antibodies such as single domain antibodies (nanobodies) and antibody-targeted nanoparticles have also been frequently explored and they seem to be promising, but these products are yet to receive approvals and reach the market. New formulation and delivery strategies are also explored with novel additives and excipients that prevent protein aggregation.

The majority of biologics, in particular mAbs, are expensive due to the high costs associated with their development and manufacturing, and thus their similar follow-on counterparts—biosimilars—have become very attractive to many consumers because of their affordability. The development of biosimilars became possible for many blockbuster biologics owing to the loss of their patent protection and updates in regulatory guidelines.

Some of the antibody products are closely related to the reference product, but because they have superior characteristics compared to the former, they are called 'biobetters'. These next-generation therapeutics are not defined well at the moment, but perhaps some existing products such as ADCs, bispecifics, and PEGylated or hyperglycosylated versions of certain products can be called biobetters. Biobetters are expected to display improved efficacy/specificity, bind to more than one target or tackle some commonly observed physical–chemical developability issues such as protein aggregation. Advances in connecting the mechanisms influencing the disposition and pharmacokinetics of mAb products will continue to augment the discovery and development of antibody products. Immunogenicity, developability and parenteral delivery remain key spaces for the optimization of antibody products as therapeutic modalities.

These new developments and other advances, such as in silico methods employed to study and/or predict protein–protein interactions, require frequent updates in the literature and necessitate the publishing of books such as this one. Herein, Siniotis et al., provide a comprehensive overview of the challenges and opportunities in the development of therapeutic mAbs. The authors highlight milestones towards antibody engineered formats, including developments in computational approaches for the strategic design of antibodies with modulated functions and an extension into novel formulation technologies such as nanocarrier delivery systems for the potential to formulate for pulmonary delivery. This overview leads into several chapters of increased detailed discussions of various mAb and mAb-based therapeutics. Leung et al., provide a detailed review of antibody drug conjugates (ADCs), which comprise of a mAb conjugated to a small molecule payload via a chemical linker and are one of fastest growing next generation mAb-based therapeutic structures. The authors cover a balance of the immense potential of ADCs to provide the promising therapeutic options in areas of unmet need along with challenges in the field to date with emphasis on antibody conjugation, linker-payload chemistry, novel payload classes, absorption, distribution, metabolism, and excretion (ADME), and product developability. The work of Voynov and colleagues, further extends and emphasizes the therapeutic tractability and clinical success of T-cell engaging bispecific antibodies as an additional high potential mAb-based biologic designed with multiple functionalities. Hotinger and May expand on and stimulate thought around the more recent application of antibody therapeutics as another approach to combat bacterial based infections due to the striking rise in resistance to antibiotics. The last two chapters of the book discuss the pragmatic challenges and considerations with mAb and mAb-based therapy developability and drug-ability. The studies reported by Sharma and coworkers share methods leveraged to test the stability of the co-formulated antibodies that can support future efforts towards the formulation and characterization of multiple high-concentration antibodies for subcutaneous delivery. In the last chapter, Boune et al., detail the importance of the consideration of N-linked glycosylation variations, which can highly influence the desired therapeutic mAb pharmacokinetics and functional properties, thereby impacting their safety and efficacy profiles.

We strongly feel this is a timely commitment and would like to express our gratitude to researchers from both academia and industry for submitting their novel mAb therapy work for this issue in order to capture the new developments that have transpired, as well as covering established concepts in this exciting field. We would like to acknowledge the authors for contributing to this timely and excellent book and the editorial office of *Antibodies* for bringing it to fruition.

Veysel Kayser, Amita Datta-Mannan
Editors

Review

Current Advancements in Addressing Key Challenges of Therapeutic Antibody Design, Manufacture, and Formulation

Vicki Sifniotis, Esteban Cruz, Barbaros Eroglu and Veysel Kayser *

School of Pharmacy, Faculty of Medicine and Health, The University of Sydney, Sydney 2006, Australia; vsif0221@uni.sydney.edu.au (V.S.); ecru7298@uni.sydney.edu.au (E.C.); barbaros.eroglu@sydney.edu.au (B.E.)

* Correspondence: veysel.kayser@sydney.edu.au; Tel.: +61-2-9351-3391

Received: 16 April 2019; Accepted: 31 May 2019; Published: 3 June 2019

Abstract: Therapeutic antibody technology heavily dominates the biologics market and continues to present as a significant industrial interest in developing novel and improved antibody treatment strategies. Many noteworthy advancements in the last decades have propelled the success of antibody development; however, there are still opportunities for improvement. In considering such interest to develop antibody therapies, this review summarizes the array of challenges and considerations faced in the design, manufacture, and formulation of therapeutic antibodies, such as stability, bioavailability and immunological engagement. We discuss the advancement of technologies that address these challenges, highlighting key antibody engineered formats that have been adapted. Furthermore, we examine the implication of novel formulation technologies such as nanocarrier delivery systems for the potential to formulate for pulmonary delivery. Finally, we comprehensively discuss developments in computational approaches for the strategic design of antibodies with modulated functions.

Keywords: therapeutic antibody; stability; aggregation; manufacture challenges; formulation

1. Introduction

Since the first therapeutic monoclonal antibody (mAb) Orthoclone OKT3® (Janssen Biotech, Horsham, PA, USA) was approved by the USA Food and Drug Administration in 1986, whole antibody therapeutics have become and persistently remain the most dominant and significant biologic therapeutic platform in the pharmaceutical industry [1,2]. To date, therapeutic antibodies treat a plethora of indications including cancers, infections, autoimmune disorders, and cardiovascular and neurological diseases [3]. The whole antibody therapeutics platform is regarded as the most promising class of pharmaceutical technology to date; it is continually being applied to newly identified biological targets and implemented in many formats to produce strategically engineered next generation antibody therapeutics, otherwise termed "biobetters" [4–9]. The international ImMunoGeneTics information system® (IMGT®, Montpellier, France) database reveals that as of December 2018, 65 whole antibodies and 18 next generation fragment or recombinant fusion antibody-based therapies are approved for clinical use, with hundreds more in clinical trials expected to reach market.

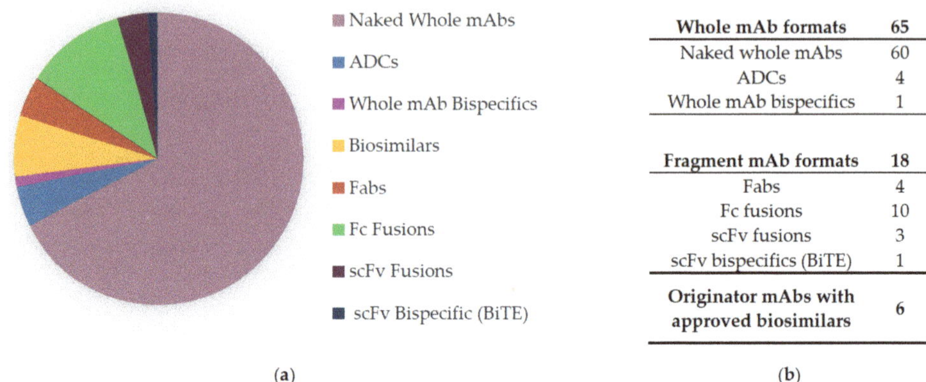

Whole mAb formats	**65**
Naked whole mAbs	60
ADCs	4
Whole mAb bispecifics	1
Fragment mAb formats	**18**
Fabs	4
Fc fusions	10
scFv fusions	3
scFv bispecifics (BiTE)	1
Originator mAbs with approved biosimilars	**6**

(b)

Figure 1. The proportions of therapeutic antibody formats approved for therapeutic use as of December 2018, IMGT® depicted through (**a**) a pie chart and (**b**) a table format.

Whole therapeutic mAbs are presently the dominant antibody platform approved for clinical use (Figure 1), although antibody engineering technologies have advanced in recent years to produce highly optimized, strategically engineered biobetter therapies, along with biosimilar mAbs reaching market to compete against their originator. Further whole mAb formats include antibody–drug conjugates (ADCs), bispecifics, isotype-switched, and glycoengineered. These additional formats have been strategically designed to introduce exceptional potency, to engage dual biological targets, and to modulate Fc effector functions. Fragments of mAbs such as the crystallizable fragment (Fc), antigen binding fragment (Fab), and single-chain variable fragment (scFv) possess key functions such as specificity to a biological target or immunological activation. The isolation of these fragments for fusion with other mAb fragments, biologically functional proteins, cytotoxic drugs, or drug carriers has been the crux of ingenuity in developing the next generation of biobetter therapies [4–15]. Figure 2 depicts several examples of prominent biobetter formats, providing a general representation of current fragment mAbs, whole mAb bispecifics, fragment mAb multispecifics, and fragment mAb fusion therapeutics.

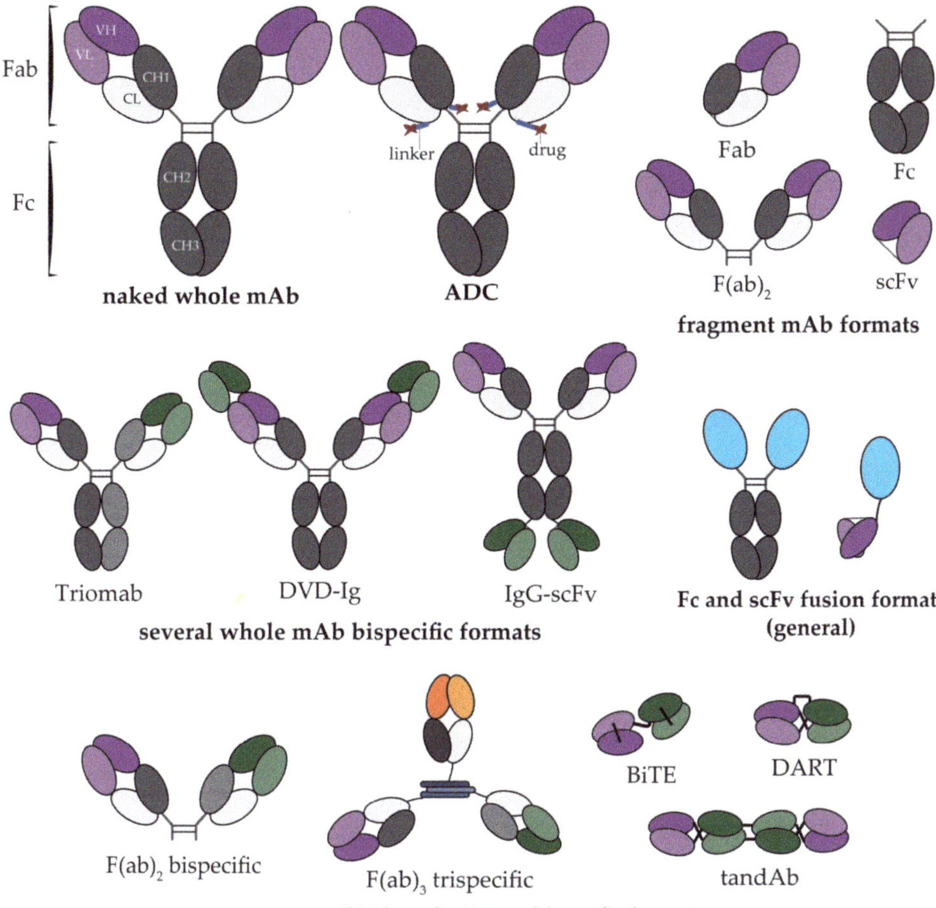

Figure 2. Schematic representation of a whole monoclonal antibody (mAb), a fragment mAb, and prominent fusion mAb formats that have been developed for strategic therapeutic uses. Proteins fused to mAb fragments are depicted as blue ovals for a general representation; however, fusion proteins may vary in size and structure. Fragment formats include the crystallizable (Fc), antigen binding (Fab and F(ab)2), and single-chain variable (scFv) fragments. Further whole mAb formats include the antibody–drug conjugate (ADC), triomab, dual variable domain immunoglobulin (DVD-Ig), and immunoglobulin–scFv fusion (IgG-scFv). Multispecific fragment formats include the F(ab)2 bispecific, bispecific T-cell engager (BiTE), dual affinity re-targeting molecule (DART), and tandem diabody (tandAb).

2. Overview of mAb Production Challenges and Considerations

Whole therapeutic mAbs require a mammalian expression system to produce the biologically functional product; however, a mAb fragment and recombinant fusion mAb products with simplified (or lacking) glycosylation are suitable for lower organism expression platforms [16]. Unlike oligopeptides, which can be chemically synthesized, whole therapeutic mAbs are considerably larger (with monomer ranging from 140–160 kDa) and comprise of four peptide chains (two heavy and two light chains) bound together by disulfide bonds and interchain non-covalent interactions. Further to this, antibodies contain glycosylation in a conserved region of the Fc (N297) that contributes to its stability and immune effector functions [17,18]. Aside from peptide synthesis, cell machinery is required to glycosylate, fold, orient, and covalently bind the antibody peptide chains in order to produce the complete, biologically functional antibody product. Manufacturing biologically functional whole mAb product is therefore commercially unfeasible through chemical synthesis and insufficient in lower organism expression platforms such as bacteria, yeast, insect, and plant cells that may not have the machinery to produce the equivalent tertiary structure and glycosylation profiles. In particular, many industrially relevant bacterial strains such as *E. coli* are completely deficient in the machinery to add post-translational glycosylations; yeasts hyper-mannosylate glycans, which cause immunogenicity; and insect cells which are deficient in sialylation machinery and produce immunogenic glycan structures [16]. A secretion of the mAb product for purification is suboptimal for several lower organism expression platforms such as *E. coli* due to poor productivity and harsh culture conditions that promote product degradation. Protein is therefore produced intracellularly, as inclusion bodies and harvest involves further processing steps such as cell lysis, inclusion body recovery, protein solubilization, and renaturation prior to further downstream purification steps [19,20]. Despite these pitfalls, the development of lower organism expression systems is of high commercial interest due to the simplified culture conditions, cheaper media requirements, rapid organism growth, and higher product yield as compared to mammalian expression systems [16,21].

In considering the requirements through the entire process of mAb discovery, manufacture, formulation, and disease treatment, several key challenges arise which have sparked overwhelming interest in pursuit of achieving better mAb manufacturing outcomes and treatment strategies. As with all biotherapeutics, mAbs and mAb-based therapeutics are limited to production in cell-based expression systems, which is considerably costly and inefficient, can have varied yields depending on the product and expression system, and requires downstream processing to remove biological contaminants introduced from the expression system. Despite affinity chromatography being a robust technology for the initial capture of a mAb for purification, the capture process and further downstream processes such as viral inactivation applies the mAb product to harsh pH and salt conditions, which can chemically degrade the mAb, leading to product instability and loss [5,22,23].

Many factors through the manufacture process influence glycosylation and charge heterogeneity of mAbs, which affects their biophysical and pharmacological properties. Though not specifically discussed in this review, the improvement and control mAb production technologies address these variations to reduce formulation heterogeneity and off-target cytotoxicities.

A common challenge, as seen with all biotherapeutics, is that mAbs and mAb-based therapeutics are currently restricted to lyophilised and liquid-based formulations for intravenous (IV) or subcutaneous (SC) delivery to achieve maximum bioavailability. Protein self-association and intrinsic stability drive this limitation, in that viscosity and propensity to aggregate are dependent on mAb concentration. Formulations are optimized to achieve the highest dosing concentration at the minimum achievable volume for injection, without compromising the quality of the mAb in formulation [24]. Viscosity remains a key limiting factor for formulating as a SC administration—certain mAb therapies are suitable and others not based on their solubility, self-association, and aggregation profiles. Alternative non-invasive administration strategies such as pulmonary delivery causes additional mechanical stress that further contribute to mAb instability and loss. Furthermore, oral delivery is unsuitable due to chemical and enzymatic degradation, as well as poor absorption in the gastric and intestinal environments [5,25–29].

A brief overview of considerations through the different concept stages of therapeutic mAb development is depicted in Figure 3. The main challenges and considerations in the manufacture and formulation of mAb therapeutics are briefly summarized in Table 1.

Figure 3. Schematic representation of the concept stages of mAb drug development in which considerations follow on from one process to the next in the design and manufacture of mAb-based therapeutics.

Table 1. Summary of key technological advancements that address challenges and considerations in mAb design, manufacture, and formulation strategies.

<table>
<tr><th>Challenges</th><th>Advancements</th></tr>
<tr><td colspan="2">Manufacture Considerations</td></tr>
<tr><td>Hybridoma technology produces immunogenic mAbs</td><td>Humanization technologies [4–6]</td></tr>
<tr><td>Yield from hybridoma technology is variable</td><td rowspan="2">Commercial cell line development and recombinant technology [4–6]; lower organism systems for fragment mAbs, mammalian systems for whole mAbs, and Fc fusions [16,30,31]</td></tr>
<tr><td>Significance of post-translational modifications and higher-order structure in mAb product</td></tr>
<tr><td>CHO expressed mAbs contain an immunogenic glycosylation profile</td><td>Human-based expression systems [16,30–32]</td></tr>
<tr><td>HEK 293 expressed mAbs are prone to aggregation</td><td>HKB-11 and PER.C6 cell lines [32]</td></tr>
<tr><td>Undesirable byproducts produced in the manufacture process</td><td>in vitro cell-free synthesis technology [33,34]</td></tr>
<tr><td>Stability of mAb affects manufacture yield due to product loss through aggregation in downstream processing steps</td><td>Enhancing mAb stability through framework mutations and hyperglycosylation [5,35–40]</td></tr>
<tr><td colspan="2">Treatment Considerations for Drug Design and Formulation</td></tr>
<tr><td>Susceptibility of mAb to degradation limits delivery to intravenous and subcutaneous only</td><td rowspan="2">Enhancing mAb stability through framework mutations, hyperglycosylation [5,35–40], and nanocarrier technologies [8,41]</td></tr>
<tr><td>Stability of mAb limits concentration of formulation</td></tr>
<tr><td>Concentration of mAb affects viscosity and injection pressure for subcutaneous delivery</td><td>Excipients, fragment mAbs, PEGylation, and hyperglycosylation [25–27,41]</td></tr>
<tr><td>Poor tissue penetration and biodistribution</td><td>Fragment mAbs and nanocarrier technologies [8,41,42]; high affinity ADCs [43–45]</td></tr>
<tr><td>Reduced half-life in low MW species</td><td>PEGylation, hyper-glycosylation and Fc fusion proteins [8,41,46]; modulation of FcRn recycling through Fc mutation [11,13]</td></tr>
<tr><td>Modulation of immunological engagement</td><td>Isotype switching, glycoengineering, and Fc mutations [11,13,15,43,46]</td></tr>
</table>

3. mAb Discovery and Manufacture Technologies

Traditional technology for whole therapeutic antibody discovery required the immunization of animals—primarily mice with a target antigen for the generation of a mixed population of B-lymphocytes-producing antibody against the target. The B-lymphocytes would be isolated for immortalization to produce monoclonal hybridoma cell lines that secrete antibody candidates of interest for further panning through display technologies to isolate potential leads [5]. This method has led to the generation of highly specific mAb libraries of non-human origin that were potentially immunogenic and less efficacious at eliciting an immune response as compared to wholly human mAbs. Technologies arose to address immunogenicity by producing chimeric and humanized antibodies through the grafting of the variable domains (chimeric) or complementarity-determining region (CDR) residues (humanized) isolated from lead non-human antibodies to a human antibody framework. To further improve the humanization strategy, transgenic mice were developed with their murine antibody heavy and light chain genes replaced with equivalent human genes; they consequently expressed wholly human antibodies for discovery [4–6].

Phage display technology remains the most prominent selection technology for panning antibody gene pools for specificity to a target antigen, as it effectively couples the expression of mAb proteins to the genes that encode them for the panning of high affinity leads. Antibody variable genes from B-lymphocyte pools are isolated through polymerase chain reaction (PCR), and cloned into a phage expression vector that presents the expressed mAb on the surface of the phage, the library of vectors is rescued in phage, and the phage library is screened against a specific target antigen. Phage pools undergo several rounds of selection to isolate high affinity candidates, the variable genes of the lead mAb candidates can then be isolated and sequenced for the design of a mAb drug format and the development of a suitable expression platform [5,47]. Error-prone PCR-based mutagenesis has propelled display technology for the generation of enormous libraries for target affinity screening. Yeast surface display further improves these screening technologies, as glycosylation sites introduced in CDRs are expressed through this platform [5,48].

The expression, yield, and quality of a mAb can vary quite substantially in hybridoma cell lines. High yielding mammalian expression systems have been developed to meet the commercial need for high expression efficiency, scalability, quality, and reproducibility. Antibody genes of interest are introduced into suitable expression vectors and transfected into highly efficient mammalian cell lines for antibody expression and secretion; a mAb can then be directly captured from the culture supernatant through affinity chromatography. Currently, the majority of mammalian expression systems for commercial whole therapeutic antibody expression are based on stable chinese hamster ovary (CHO), mouse myeloma (NS0), and mouse hybridoma (Sp2/0) cell lines [4,5]. However, in the context of research and development, the transient expression in human cell lines such as embryonic kidney (HEK 293), amniotic (CAP), a hybrid of HEK 293 and lymphoma (HKB-11), and embryonic retina (PER.C6) are favored over stable CHO expression due to the ease and speed of production of workable quantities of antibody for preliminary studies [16,30–32,49,50]. The generated mAb library undergoes relevant in vitro testing along with formulation stability screening to exclude candidates that have poor manufacturability attributes [51]. Lead mAb candidates that show potential for further investigation would then be developed for high yielding stable expression and more rigorous characterization leading up to therapeutic development. However, a drawback from changing from human to hamster expression systems is the difference in glycosylation profile. CHO expressed mAbs produce immunogenic non-human glycoform Neu5Gc and have a higher composition of sialylation, which may result in reduced antibody-dependent cellular cytotoxicity (ADCC). Therefore in the early stages of development, leads need to be identified and thoroughly characterized in the expression system to be developed for commercial manufacture before committing to industrial scale up [16,30,32]. Amongst other biotherapeutics, approved mAb Fc fusion therapeutics dulglutide (Trulicity®, Eli Lilly, Indianapolis, IN USA), efmoroctocog alpha and eftrenonacog alpha (Eloctate® and Alprolix®, Biogen, Cambridge, MA, USA), are manufactured in a HEK 293 expression system, which is giving rise to the acceptance of human-based expression systems for the production of mAb-based therapeutics [4]. However, HEK 293 expression systems are prone to inducing mAb aggregation in cultures, which is detrimental to cell viability and creates a loss of product during manufacture. Hence, HKB-11 and PER.C6 are the preferred commercial human cell lines for mAb expression in human cell lines, specifically [16,31,32].

Established lower organism expression systems for fragment or recombinant fusion mAb therapeutics include bacteria such as *E. coli*, yeasts such as *S. cerevisiae* and *P. pastoris*, plants such as tobacco, algae, and insects such as silkworm [5,16,52–54]. Expression platforms that use *E. coli* in particular are considered high risk due to the potential endotoxin contamination in the mAb product, of which complete endotoxin removal requires further purification steps [19]. However, several approved mAb fragment- and recombinant-based therapies are produced in an *E. coli*-based expression system, including pegol conjugated Fab' certolizumab pegol (Cimzia®, Celltech UCB, Brussels, Belgium), Fab ranibizumab (Lucentis®, Genentech, San Francisco, CA, USA), and recombinant Fc fusion romiplostim (Nplate®, Amgen, Thousand Oaks, CA, USA). Lower organism expression platforms such as *E. coli* and *P. pastoris* continue to be the preferred option for the manufacture of fragment mAb formats due to the relative ease, high yield, and reduced cost for manufacture as compared to mammalian expression platforms.

An emerging in vitro cell-free synthesis technology is being developed with bacterial and CHO cell lysates which have the potential to alleviate formation of undesirable biological byproducts, as the machinery from the cell lysates purely express protein from the mAb genes that are introduced to the system. Though this technology is not currently applicable for industrial scale manufacture, it holds much promise as an alternative to live culture. For instance, culture maintenance and highly defined media are no longer necessary, reactions can run continuously, lysates can be recycled with reproducible results, unmodified linear DNA is suitable for the system (thus alleviating a need for multiple cloning steps), and additional enzymes can be introduced to the system for the engineering of specific post-translational modifications [33,34,55–58]. Despite the apparent advantages of this technology for mAb manufacture, cell lysates encounter the same challenges as their host cell expression system. That is, post-translational modifications of mAb product from lysates from lower level organisms are limited by the endogenous machinery of that host organism. Furthermore, preparation of mammalian cell lysates is challenging, and yield produced from these lysates is considerably low [59,60].

Current bioprocess engineering and cell line development strategies have been crucial in enhancing the manufacturability of mAbs. Many current mAb therapeutic expression platforms make use of CHO cell lines that have been commercially developed with dihydrofolate reductase or glutamine synthetase deficiency for enhanced stability selection through resistance to methotrexate and methionine sulfoximine inhibition, respectively [50,61]. Mammalian cell lines have been adapted to suspension culture, as they support higher cell densities and mAb titers in the absence of serum from the culture media, for large scale fed-batches, perfusion systems, or continuous culture systems. Further to this, cell lines are continually being engineered for enhanced metabolic functions, introducing glycosylation pathways, superior secretion, and resistance to apoptosis for prolonged survival, in efforts to generate super-producer cell lines that can sustain a continuous culture [62–68].

Gene and expression vector design and development are specifically tailored to the expression system to maximize mAb expression and cell line stabilization through host cell codon optimization and the addition of highly efficient transcription, secretion, selection, and integration elements [69–73]. Vectors can contain a single site for gene insertion for the subsequent co-transfection of vectors that carry the antibody heavy and light chains, or both chains can be cloned into a dual expression vector. In more elaborately designed recombinant mAb drug formats that require the expression of several genes, vector technologies have implemented multicistronic expression to enhance the efficiency of the vector system. The stable or transient transfection of vectors is performed on highly viable cell cultures with high cell density. Antibody heavy and light chain ratios may require adjustment in transfection for optimizing expression and secretion. A reporter vector that expresses green fluorescent protein is typically included to observe transfection efficiency during optimization [74,75]. Liposome-mediated transfection is preferential over all other mammalian transfection strategies and is induced chemically through the complexing of DNA to a cationic lipid prior to or during addition to a culture. Polyethylenimine is the most prevalently used transfection reagent despite the development of superior reagents such as Lipofectamine™ 2000 and Freestyle™ MAX (Thermo Fisher Scientific, Waltham, MA, USA), LyoVec™ (InvivoGen, San Diego, CA, USA), FuGENE 6™ (Promega, Madison, WI, USA), and *Trans*IT-PRO® (Mirus, Madison, WI, USA), owing to its lower cost and relative efficiency [73,75–78].

The commercial manufacture of mAbs has transitioned to serum-free, chemically defined, and animal-free media which has removed a source of expression variability and has been driven particularly from the advent of mad cow disease [50,79]. Several media supplements such as plant and yeast digests (peptones/hydrolysates), surfactants (e.g., Pluronic F-68), DNA methyltransferase (azacytidine), and histone deacetylase (sodium butyrate, valproic acid) inhibitors have been found to support cell viability and enhance mAb expression [79–85]. Mild hypothermia of the culture has also demonstrated to reduce cell expansion, support prolonged cell viability, and enhance mAb expression [73,86–90]. Further to this, continuous supplementation in expression cultures of uridine, manganese chloride, and galactose demonstrated successful the glycan engineering of expressed mAb, where mAb hyper-galactosylation was promoted to enhance complement-dependent cytotoxicity (CDC) activity [91].

In the manufacturing process of fed-batch, perfusion, and continuous culture systems, media is continuously fed in the culture system, whilst parameters such as cell viability, cell density, and metabolite levels are monitored in real time until the parameters indicate that it is the optimal time to harvest the expressed mAb from the culture supernatant. In perfusion and continuous culture systems, a feed is removed from the culture simultaneously to the media addition, the difference being that the cell dilution rate in continuous culture is optimized to remain equal or higher than the cell growth rate, which allows the perpetuation of mAb expression and requires the continuous harvest of the supernatant [92–94]. The harvest of the supernatant requires clarification technologies such as the use of precipitants/flocculants (e.g., polyethylene glycol, diethylaminoethyl dextran, caprylic acid, and polyethylenimine), high throughput centrifugation, and filtration methods to remove cells and biological debris, for the lowering the loading burden in the lead up to mAb capture through affinity chromatography [19,95–100]. The mechanical stresses from centrifugation and filtration are unavoidable, although they can induce mild mAb product loss through fragmentation and aggregation.

Protein A-ligand-based affinity chromatography is the most robust, efficient, and prevalently used mAb capture technology, owing to its selectivity and high affinity to various human Fc, as well as its efficient dissociation of captured mAbs at a low pH for reuse. Fragment mAbs and recombinant formats based on Fabs and single chain variable fragments (scFv) are unable to take advantage of protein A affinity capture, and alternatives have been developed, such as capturing proteins G, M, and L, which bind at different mAb epitopes; ion exchange chromatography; and polyhistidine tagged capture through immobilized metal chromatography [100–102]. A further advantage of affinity chromatography in manufacture is the integration of a viral inactivation step by holding the low pH-eluted mAb product prior to further purification steps; however, this applies the mAb product to low pH at high concentrations, which can induce mild mAb instability, aggregation, and the formation of acidic variants [5,22,100,103,104].

Post mAb capture, further polishing steps are required to remove further contaminants such as host cell proteins and DNA, as well as leached affinity chromatography ligand- and mAb-degradation products such as fragments, aggregates, and ionic variants. The polishing steps are designed based on the specific properties of the mAb product, such as the isoelectric point and molecular weight (MW), to implement appropriate chromatographic technologies such as anion and cation exchanges, hydrophobic interaction, and multimodal and size exclusion that will separate the mAb product from the manufacture-introduced impurities [22,100,103]. The additional chromatographic steps unavoidably apply the mAb product to buffers of varying ionic strength and high concentrations, which can again induce mild mAb instability and aggregation. Post polishing steps, the purified mAb product undergoes a further viral removal step, such as filtration, and then it undergoes buffer exchange and concentration through ultrafiltration or diafiltration methods to prepare the bulk mAb product for formulation [100,105].

4. Formulation Strategies and Considerations

Strategies to improve the formulation of mAbs and mAb-based therapies continues to be an ongoing challenge, as is faced with all biotherapeutics. Firstly, whole therapeutic mAbs are unsuitable for non-invasive oral, nasal, or pulmonary routes of administration as they are susceptible to chemical and enzymatic degradation in the gastrointestinal tract. The bioavailability of mAbs through these routes is poor, owing to mAbs' polar surface charge and relatively large MW, limiting transport through mucosal membranes.

Fragment mAb platforms, together with excipient and PEGylation technologies, have been developed to help circumvent transportation limitations for pulmonary delivery, specifically. However, physical stresses applied to the mAb (e.g., shear stresses from the aerosolization of liquid formulations for a pressurized meter dose and a nebulization delivery, or from the production of dry powder for inhalation) pose further challenges of mAb instability, leading to degradation and reduced efficacy. [8,26,28]. Few biologics have been successfully approved for pulmonary delivery, most notably insulin formulation Afrezza® (MannKind, Westlake Village, CA, USA), and, currently, an erythropoietin-Fc fusion and a nanobody-targeting respiratory syncytial virus are in clinical trials, showing promise for the further development of this administration strategy for both the systemic and localized delivery of mAb-based therapeutics [5,28].

Several drug carrier technologies such as microencapsulation, liposome, and nanoparticle formulations inherently enhance the stability and control the release of mAbs, which can prolong their half-life. These nanocarrier formulation strategies are of intense interest, as they hold promise for developing less invasive inhaled formulations of mAb-based therapeutics with superior attributes to currently established formulations [41,106–109].

The majority of currently approved mAb therapeutics are formulated for IV, although commercial interest has directed technologies to develop injectable mAb formulations which has seen success in many approved mAb therapies thus far, primarily SC for systemic delivery, along with a few intravitreal and intramuscular formulations for tissue-specific indications. Amongst others, anti-HER2 antibody trastuzumab was originally developed as an IV formulation and was successfully repurposed as a SC formulation [110–112], whereas next generation therapies have directly moved towards SC formulation, such as with anti-TNF-α antibody adalimumab [113]. Injectable administrations offer several advantages over IV administration, especially in the treatment of chronic diseases in regards to a reduced burden to allied health services, patient tolerance, and adherence to treatment; however, the intrinsic physicochemical properties of mAbs may be undesirable for injectable formulation. In considering the low volume, injection pressure, and the typically high (>100 mg) effective dose of mAb for injectable delivery, the viscosity and aggregation propensity of mAbs become key formulation challenges to address, as they are dependent on mAb concentration. Certain mAb therapies are suitable, and others are not based on their solubility, viscosity, self-association, intrinsic stability, aggregation, and precipitation profiles.

Injectable mAb formulations can be further improved with the use of excipients to increase solubility, reduce viscosity, and enhance the stability of mAbs. Excipients are considered for a formulation based on their physicochemical properties, pharmacokinetics, and safety. For example, polysorbates are commonly used in biologics as a stabilizing agent; however, their addition in high concentrations can denature proteins and cause adverse side effects such as injection site reactions [114–116]. Injectable mAb formulations are co-formulated with recombinant human hyaluronidase, specifically as a permeation enhancer for more efficient absorption into tissue, although the inclusion of this additional biologic adds further burden to the formulation's viscosity and propensity to aggregate [5,8,25–28,117–123]. Antibody therapies for IV administration are prepared as lyophilised powder for reconstitution and further dilution, and injectable administrations are prepared as liquid-based formulations in pre-filled syringes. Liquid formulations of mAbs are more susceptible to physiochemical degradation, are less stable, and have a reduced shelf-life as compared to lyophilised formulations. However, drying technologies to produce lyophilised formulations apply the mAb to physical stresses that induce instability and degradation, leading to reduced efficacy [124–126].

Long-term stability predictions are elucidated from formulation screening in accelerated aggregation studies, as part of the preliminary screening process of generated mAb libraries [127]. The stability profiles of mAbs can be greatly affected by the amino acid composition, structure, and potential glycosylation in their variable region CDRs that are isolated through the screening and maturation process. Elucidating the causes for reduced solubility or increased aggregation propensity has to be considered together with the molecular interactions between the mAb and the biological target as to not compromise affinity. Computational tools have been developed to elucidate amino acids within the mAb CDR structure that have a high propensity to aggregate, and strategies for improving solubility and aggregation profiles have been developed through direct amino acid substitutions and the strategic addition of glycosylation sites [128–132]. The characterization of mAb stability and target interaction for optimization is considered a fundamental step as part of preliminary drug discovery, as the applicability for development, manufacture, and formulation of the mAb therapeutic is governed by the mAbs stability profile.

5. Improving mAb Tissue Penetration for Cancer Treatment

Antibody therapies are currently restricted to invasive parenteral routes of administration for maximum bioavailability and systemic distribution; however, delivery to the specific target tissue, such as tumors for cancer treatments, continues to be a challenge. Penetration to tissue from blood vessels is again poor, owing to mAbs' polar surface charge and relatively large MW, limiting transport through physiological barriers. The efficiency of tissue penetration is further influenced by systemic and local mAb clearance rates. Enhancement to mAbs' affinity to their biological target through maturation and strategic mutation technologies have led to faster diffusion rates of mAb to target for increased efficacy. However, for tumourous tissue specifically, penetration through tumors is restricted by the tumor's binding site barrier, in which antigens expressed on the tumors periphery capture the majority of the mAb released to surrounding tissue; this effect is exaggerated with overexpressed antigen. The enhancement of whole mAbs' affinity beyond 1 nM has specifically shown that there is no further improvement of tumor diffusion rates, tissue penetration, and accumulation [42,43,133]. An increased affinity of mAbs to cellular targets also leads to the increased uptake, internalization, and catabolism of mAbs, which reduces ADCC and increases the mAb clearance rate. This mechanism is exploited through the development of high affinity ADCs to target the delivery and release of potent drugs, inducing targeted cell death [43–45].

Fragment mAb platforms have shown better tissue penetration and biodistribution than whole mAb therapeutics; however, a pitfall of smaller peptides lacking an Fc region is a highly reduced in vivo half-life and poor retention times. Technologies to improve the half-life of mAb fragments have been primarily through PEGylation, along with strategic Fc mutation and glycosylation engineering to enhance neonatal Fc receptor (FcRn) recycling. Furthermore, hyper-glycosylation technology has been successfully applied in other biotherapeutics, such as with glycosylated erythropoietin Aranesp® (Amgen, Thousand Oaks, CA, USA). Conversely, nanocarrier platforms that are relatively larger as compared to whole mAbs have demonstrated superior pharmacokinetics and tumor retention, as well as previously mentioned enhanced stability and the sustained, controlled release of mAbs [41]. These enhanced properties have generated considerable interest in developing the systemic delivery of mAb-nanoparticle platforms for superior tumor penetration, as well as targeted and sustained therapeutic drug delivery. Further to nanocarriers, additional formulation strategies developed for the controlled release of mAbs include hydrogels and crystalline antibodies, which have shown success as stable, injectable formulations for development [5,8,15,41–43,134–136].

In addition, to further enhance the stability and half-life of other biotherapeutics, recombinant technologies allowed their fusion to mAb Fc as to introduce FcRn recycling as a protection mechanism, which has innovatively expanded the applicability of mAb-based therapeutic platforms. Notable examples include the TNF Receptor–Fc fusion protein Enbrel® (Amgen, Thousand Oaks, CA, USA), which acts as an inhibitor to overexpressed TNF-α in autoimmune diseases, and the Factor IX–Fc fusion protein Alprolix® (Biogen, Cambridge, MA, USA), which is a blood factor supplement for hemophiliacs [11,46,133].

6. Strategic Modulation of mAb Immune Effector Functions

The modulation of mAbs effector functions through isotype switching, glycoengineering, and strategic mutations have proved advantageous for the development of more effective mAb treatment strategies. Antibody binding to Cq1 promotes the complement cascade, FcγR1A/B, FcγR2A, and FcγR3A/B receptors to activate immune effector functions; FcγR2B counter-balances the effector response, and FcRn prolongs mAb half-life. Binding to these receptors is primarily done through sites in the C_{hinge}, C_H2, and the conserved glycosylation region in the Fc (N297).

IgG3 does not efficiently bind to FcRn, which reduces its half-life to approximately seven days, as opposed to a 21 day half-life for the other IgG isotypes. In most instances of mAb design, an extended half-life is preferred as to prolong the effective dose of a mAb in serum [15]. IgG2 and IgG4 mAbs have a reduced effector function as compared to IgG1 and are thus used in instances where minimal engagement to the immune system is warranted to increase safety of the mAb therapy—with ADCs reducing off-target cytotoxicity, for instance. Eculizumab (Soliris®, Alexion Pharmaceuticals, New Haven, CT, USA) is the first (and thus far only) approved recombinant IgGκ 2(C_H1/C_{hinge})–4(C_H2/C_H3) hybrid whole mAb. Further cross-sub-class variant mAb therapies are in development, with key amino acid substitutions from the IgG2 and IgG4 sub-classes introduced for intentionally suppressed effector functions [4,13]. The removal of the conserved glycosylation site through amino acid substitution of N297 or T299 (aglycosylation) has also demonstrated reduced effector function; however, this happens at the expense of mAb stability and is therefore not a suitable strategy for whole mAb therapeutics [11,14,15,137–140].

On the other hand, enhancing ADCC and CDC is useful for engaging the immune system to tumor tissues in the absence of a conjugated cytotoxic drug. Certain glycoengineered modifications to the conserved glycosylation region in the Fc, such as deficiency in core fucose (afucosylation) and hyper-galactosylation, have demonstrated enhanced mAb binding to FcγR3A and Cq1, specifically for an enhanced ADCC and CDC effect [13,43]. Afucosylated mAbs benralizumab (Fasenra™, AstraZeneca, London, UK) and mogamulizumab (Poteligeo®, Kyowa Hakko Kirin, Tokyo, Japan), as well as low fucose content mAb obinutuzumab (Gazyva®, Genentech, San Francisco, CA, USA), are currently approved therapies, with several further in development (along with clinical trials), which demonstrates the commercial interest and applicability of this technology in improving mAb therapeutics through enhancing ADCC. However, the significance of manipulating sialylation in mAb therapeutics is a controversial topic, with several reports demonstrating a reduction in ADCC and CDC and others reporting no observable difference. This highlights a clear need to characterize and define the in vivo efficacy of such manipulation [11,13,14,43].

No currently approved mAb therapies contain amino acid substitutions for an improved half-life through modulated binding properties to FcRn, improved CDC through binding to complement factor Cq1, or improved ADCC through enhanced binding affinity to FcγR1A and 3A. However, several mutations have been reported, patented, and are in clinical development, demonstrating the commercial interest in this technology for improving mAb therapeutics [6,11,13,15,133,137,140–143].

For a more comprehensive understanding of mAb engineering strategies for modulated immune effector functions, we direct the reader to the following reviews [13–15,140].

7. Computational Approaches for Aggregation Prediction and Rational Design of mAbs

The advancement of in silico analysis of mAb peptide sequences, structures, conformation, and their associated biological interaction has been integral to the development of various computational tools for characterizing, designing, and optimizing mAb-based therapeutics. The generation and continued pursuit of mAb structural data for in silico analysis has led to the development of several integral databases, notably IMGT®, serving as a key resource for data mining [144–146]. Molecular dynamic (MD) simulation analysis has driven the computational analysis of the discovery and characterization of relevant molecular interactions within the mAb molecule, mAb binding to biological targets and mAb surface association with the surrounding environment. These interactions can infer intrinsic stability, target binding associations, solubility and aggregation propensity of the mAb. MD simulations and free energy calculations from crystal structures remain key for the highly specific elucidation of mAb-target intermolecular interactions that correlate to binding affinity, as significant interactions such as hydrogen-bond formation can be predicted and determine the strength of the molecular associations [147]. However, elucidation of mAb self-association, solubility, and aggregation propensity have been driven by the development of computational modelling and simulation tools that simultaneously analyses mAb topography and surface polarity. Notably, AGGRESCAN3D, TANGO, and PASTA are the most prominent tools for predicting the site specific aggregation propensity of mAbs [35,148].

The preliminary elucidation of mAb structural data—that being amino acid sequences and higher order structure (HOS)—is experimentally derived to produce crystal structures that model the solid-state 3D structure of the mAb. Mass spectrometry technologies have come of age to produce high throughput and orthogonal analysis to elucidate mAb peptide sequences, oxidation, deamidation, and glycosylation heterogeneity, as well as, more recently, for native, destabilized, and aggregated HOS elucidation [149,150]. X-ray crystallography technologies have been the underlying workhorse for elucidating the crystal structure of mAbs. Producing crystalline mAbs is challenging, owing to the complexity of mAb HOS and the degree of mAb conformational heterogeneity. However, several complementing technologies have evolved in recent years to ascertain structure and interactions, including circular dichroism (CD), infrared (IR) and raman spectroscopy, cryogenic-electron microscopy, and nuclear magnetic resonance (NMR) spectroscopy [148,150–157]. Of the technologies available for HOS elucidation, 2D-NMR and X-Ray crystallography (followed by MS) provide the highest sensitivity and local specificity. In comparison, CD, IR, and raman are high-throughput methods, although they have much lower sensitivity [150,157].

Thus far, only four whole IgG antibodies have a successfully determined crystal structure (PDB ID: 1HZH, 1IGT, 1IGY, and 5DK3), notably 1HZH as a wholly human IgG1 and 5DK3 as a humanized IgG4/κ, both of which have been used as model structures for MD simulations [158]. However, fragment mAbs yield much higher success with producing crystal structures, which has led to much pursuit and contribution in generating mAb fragments and targetting complexing crystal structure libraries for data mining and in silico analysis [159,160].

Upon obtaining relevant crystal structures for a candidate mAb, detailed crystal structure and MD simulation analysis has been extensively used to elucidate the mAbs intramolecular interactions that confer intrinsic stability and intermolecular interactions to confer interactions to biological targets and self-association. The particular binding interface of interest is analyzed, and amino acids in the mAb structure are identified for substitution that can disrupt molecular interactions in the interface in the instance of diminishing target binding and identify potentially unutilized bonding sites and polarity mismatches in the interface that can be improved in the instance of enhancing target binding. MD simulations are then carried out for the native and substituted mAb structures to elucidate any relevant bond formations, interactions, and more favorable binding free energy produced by the substituted structure, of which promising candidates can be synthesized and validated through in vitro analysis. This predictive approach has been successfully applied to identify mutations in mAbs for modulating effector functions, target binding, optimizing affinity capture for manufacturing, and, more recently, for designing antibodies de novo from targets of interest [11,161–165]. Mutations have been specifically identified for enhanced binding to FcγR2A, FcγR3A, and Cq1, as well as a reduced binding to FcγR2B for a more pronounced ADCC and CDC effect, an enhanced binding to FcRn for an extended half-life, and a reduced binding to FcγRs and Cq1 for a diminished ADCC and CDC effect, all of which are transferrable to mAbs of the same isotype sub-class [137,141,142]. Further to this, the molecular interactions of glycoengineered mAb variants to FcRs have been characterized through MD simulations to corroborate predictions with the observed modulated effector functions [13,15,137,141,142].

Self-association, solubility, and aggregation propensity are further evaluated by computational modelling tools that specifically characterize the topography and surface polarity of mAbs, concurrently analyzing the local spatial arrangement of amino acids in the mAb structure, solvent accessibility, and local surface charge [5]. In particular, surface exposed hydrophobicity is sought as the lead mechanism for protein self-association driving aggregation propensity, in which aggregation-prone regions (APRs) are identified in the mAb structure. The substitution of identified APRs to gatekeeper (i.e., polar) amino acids has experimentally demonstrated resistance to aggregation and improved solubility, which validates this strategy for improving mAb stability [36–40,166–170]. The majority of APRs identified are located in biologically relevant mAb regions—those being the CDRs and the Fc. Specifically targeting these APRs through mutation may disrupt important biological functions and mAb structure. Analyses of the mAbs intermolecular interactions are necessary to validate the intrinsic stability of substituted mAb variants; if conformational fluctuations are elucidated, then the biologically active interface is likely disrupted. Interestingly, mutations for enhancing biological functions have demonstrated a negative impact on mAb solubility and stability, further suggesting that the self-association profile of mAbs is linked to its biological activity [36,128,171].

Aside from direct residue substitution in mAb structures, other strategies reported to profoundly interfere with the effects of APRs include isotype switching and the strategic addition of N-linked glycans [39,128,172,173]. Additionally, N-linked glycosylation in the CDR regions of mAbs, although uncommon, has demonstrated improved solubility and stability profiles of mAbs without impacting their target affinity, as compared to the same mAbs with removed CDR glycans [128,174]. The rationale behind strategic glycan addition is based on the election of N-linked glycosylation sites that have apparent spacial proximity to APRs, with the introduced glycan therefore sterically hindering the APR from self-association interactions. The benefits of extending mAb half-life with strategic glycan addition, as seen with hyper-glycosylated biotherapeutics (as well as improving mAb solubility and stability for improved formulation strategies) has yet to be realized and suggests a very intriguing and highly relevant technology for perusal.

8. Concluding Remarks

Therapeutic antibodies have come of age as continuing to be a key, dominant technology in the biopharmaceutical industry. The repurpose of antibodies to many formats have made them versatile to design and tailor highly specialized treatments, including ADCs as a targeted drug delivery system, bispecific and fragment mAb platforms for tailored engagement and increased bioavailability, and recombinant Fc-fusion proteins for an increased half-life and introduced immunological engagement. Further advancements include the modulation of Fc effector functions through manipulations of the Fc, either through isotype switching, glycoengineering, or strategic mutations in the Fc region, along with the PEGylation of fragment mAbs for enhanced half-life. Advancements in discovery, manufacture, and formulation technologies have further propelled the success of therapeutic antibodies, notably through expression system development and the transition from IV to SC formulations. Human-based expression systems have been extensively used in mAb development and are becoming an accepted manufacturing platform for mAb therapeutics. Furthermore, cell-free synthesis technology is giving rise to the potential for higher efficiency in the manufacture process.

Though developments for further generation antibody therapies have led to great strides in producing improved therapeutic outcomes, many facets of the manufacture process and formulation development strategy pose as challenges to be considered. Antibody-based therapies are susceptible to chemical and enzymatic degradation through oral, nasal, or pulmonary routes of administration and are therefore currently restricted IV or SC delivery. Despite achieving maximum bioavailability through IV/SC administration, tissue penetration of mAb-based therapies is poor, which limits their local bioavailability, requiring high concentrations to achieve an effective dose. The stability of mAbs-based therapeutics is a highly pronounced and recurring challenge to be considered, as it affects manufacture yield and formulation considerations. Several strategies are in development to improve the stability of mAbs in order to potentially produce formulations for pulmonary or oral administration. Notably, computational tools have come of age, complementing experimental techniques to derive antibody structure and aggregation prediction. Through these methods, the stability and aggregation propensity of mAb-based therapies have demonstrated improvement through rational mutation and glycosylation within the framework region, which is potentially translatable to all mAbs within the same isotype. Furthermore, nanocarrier technologies have been shown to enhance the stability and potentially control the release of mAbs. The refinement of rational mAb design coupled with nanocarrier technologies has the potential to overcome these challenges, to develop superior treatment strategies, and ultimately to formulate for non-invasive administration routes such as pulmonary delivery.

Author Contributions: V.S. revised the figure and prepared the scope, literature review, and original manuscript draft. E.C prepared the original figure and reviewed the manuscript. B.E. and V.K. equally reviewed and edited the manuscript.

Funding: This research received no external funding.

Conflicts of Interest: The authors declare no conflict of interest.

References

1. Urquhart, L. Top drugs and companies by sales in 2017. *Nat. Rev. Drug Discov.* **2018**, *17*, 232. [CrossRef] [PubMed]
2. Ecker, D.; Dana Jones, S.; Levine, H.L. The therapeutic monoclonal antibody market. *mAbs* **2015**, *7*, 9–14. [CrossRef] [PubMed]
3. An, Z. "Magic bullets" at the center stage of immune therapy: A special issue on therapeutic antibodies. *Protein Cell* **2018**, *9*, 1–2. [CrossRef] [PubMed]
4. Santos, M.L.d.; Quintilio, W.; Manieri, T.M.; Tsuruta, L.R.; Moro, A.M. Advances and challenges in therapeutic monoclonal antibodies drug development. *Bras. J. Pharm. Sci.* **2018**, *54*. [CrossRef]
5. Elgundi, Z.; Reslan, M.; Cruz, E.; Sifniotis, V.; Kayser, V. The state-of-play and future of antibody therapeutics. *Adv. Drug Del. Rev.* **2017**, *122*, 2–19. [CrossRef]
6. Almagro, J.C.; Daniels-Wells, T.R.; Perez-Tapia, S.M.; Penichet, M.L. Progress and challenges in the design and clinical development of antibodies for cancer therapy. *Front. Immunol.* **2018**, *8*, 1751. [CrossRef] [PubMed]
7. Liu, B.N.; Luo, J.H. Research and development of innovative antibody-based drugs. *Yaoxue Xuebao* **2017**, *52*, 1811–1819.
8. Awwad, S.; Angkawinitwong, U. Overview of antibody drug delivery. *Pharmaceutics* **2018**, *10*, 83. [CrossRef] [PubMed]
9. Klein, C. Special issue: Monoclonal antibodies. *Antibodies* **2018**, *7*, 17. [CrossRef]
10. Wang, C.; Xu, P.; Zhang, L.; Huang, J.; Zhu, K.; Luo, C. Current strategies and applications for precision drug design. *Front. Pharmacol.* **2018**, *9*, 787. [CrossRef] [PubMed]
11. Strohl, W.R. Current progress in innovative engineered antibodies. *Protein Cell* **2018**, *9*, 86–120. [CrossRef] [PubMed]
12. Hoffmann, R.M.; Coumbe, B.G.T.; Josephs, D.H.; Mele, S.; Ilieva, K.M.; Cheung, A.; Tutt, A.N.; Spicer, J.F.; Thurston, D.E.; Crescioli, S.; et al. Antibody structure and engineering considerations for the design and function of antibody drug conjugates (adcs). *OncoImmunology* **2018**, *7*, e1395127. [CrossRef] [PubMed]
13. Wang, X.; Mathieu, M.; Brezski, R.J. Igg fc engineering to modulate antibody effector functions. *Protein Cell* **2018**, *9*, 63–73. [CrossRef] [PubMed]

14. Pawlowski, J.W.; Bajardi-Taccioli, A.; Houde, D.; Feschenko, M.; Carlage, T.; Kaltashov, I.A. Influence of glycan modification on igg1 biochemical and biophysical properties. *J. Pharm. Biomed. Anal.* **2018**, *151*, 133–144. [CrossRef] [PubMed]
15. Fonseca, M.H.G.; Furtado, G.P.; Bezerra, M.R.L.; Pontes, L.Q.; Fernandes, C.F.C. Boosting half-life and effector functions of therapeutic antibodies by fc-engineering: An interaction-function review. *Int. J. Biol. Macromol.* **2018**, *119*, 306–311. [CrossRef] [PubMed]
16. Mizukami, A.; Caron, A.L.; Picanço-Castro, V.; Swiech, K. Platforms for recombinant therapeutic glycoprotein production. In *Recombinant Glycoprotein Production*; Humana Press Inc.: New York, NY, USA, 2018; Volume 1674, pp. 1–14.
17. Thomson, C.A. Igg structure and function. In *Encyclopedia of Immunobiology*; Elsevier Inc.: Amsterdam, The Netherlands, 2016; Volume 2, pp. 15–22.
18. Kayser, V.; Chennamsetty, N.; Voynov, V.; Forrer, K.; Helk, B.; Trout, B.L. Glycosylation influences on the aggregation propensity of therapeutic monoclonal antibodies. *Biotechnol. J.* **2011**, *6*, 38–44. [CrossRef]
19. Pieracci, J.P.; Armando, J.W.; Westoby, M.; Thommes, J. Chapter 9—Industry review of cell separation and product harvesting methods. In *Biopharmaceutical Processing*; Jagschies, G., Lindskog, E., Łącki, K., Galliher, P., Eds.; Elsevier: Amsterdam, The Netherlands, 2018; pp. 165–206.
20. Huang, C.-J.; Lin, H.; Yang, X. Industrial production of recombinant therapeutics in escherichia coli and its recent advancements. *J. Ind. Microbiol. Biotechnol.* **2012**, *39*, 383–399. [CrossRef]
21. Behme, S. *Manufacturing of Pharmaceutical Proteins: From Technology to Economy: Second, Revised and Expanded Edition*; Wiley: Hoboken, NJ, USA, 2015; pp. 1–427.
22. Ulmer, N.; Vogg, S.; Müller-Späth, T.; Morbidelli, M. Chapter 7—Purification of human monoclonal antibodies and their fragments. In *Human monoclonal Antibodies: Methods and Protocols*; Steinitz, M., Ed.; Springer: New York, NY, USA, 2019; pp. 163–188.
23. Arosio, P.; Barolo, G.; Müller-Späth, T.; Wu, H.; Morbidelli, M. Aggregation stability of a monoclonal antibody during downstream processing. *Pharm. Res.* **2011**, *28*, 1884–1894. [CrossRef]
24. Kayser, V.; Chennamsetty, N.; Voynov, V.; Helk, B.; Trout, B.L. Conformational stability and aggregation of therapeutic monoclonal antibodies studied with ans and thioflavin t binding. *mAbs* **2011**, *3*, 408–411. [CrossRef]
25. Bittner, B.; Richter, W.; Schmidt, J. Subcutaneous administration of biotherapeutics: An overview of current challenges and opportunities. *Biodrugs* **2018**, *32*, 425–440. [CrossRef]
26. Viola, M.; Sequeira, J.; Seiça, R.; Veiga, F.; Serra, J.; Santos, A.C.; Ribeiro, A.J. Subcutaneous delivery of monoclonal antibodies: How do we get there? *J. Control. Release* **2018**, *286*, 301–314. [CrossRef] [PubMed]
27. Otvos, L. 6.05—Basic principles of formulation for biotherapeutics: Approaches to alternative drug delivery. In *Comprehensive Medicinal Chemistry III*; Chackalamannil, S., Rotella, D., Ward, S.E., Eds.; Elsevier: Oxford, UK, 2017; pp. 131–156.
28. Morales, J.O.; Fathe, K.R.; Brunaugh, A.; Ferrati, S.; Li, S.; Montenegro-Nicolini, M.; Mousavikhamene, Z.; McConville, J.T.; Prausnitz, M.R.; Smyth, H.D.C. Challenges and future prospects for the delivery of biologics: Oral mucosal, pulmonary, and transdermal routes. *AAPS J.* **2017**, *19*, 652–668. [CrossRef] [PubMed]
29. Muheem, A.; Shakeel, F.; Jahangir, M.A.; Anwar, M.; Mallick, N.; Jain, G.K.; Warsi, M.H.; Ahmad, F.J. A review on the strategies for oral delivery of proteins and peptides and their clinical perspectives. *Saudi Pharm. J.* **2016**, *24*, 413–428. [CrossRef] [PubMed]
30. Hunter, M.; Yuan, P.; Vavilala, D.; Fox, M. Optimization of protein expression in mammalian cells. *Curr. Protoc. Protein Sci.* **2019**, *95*, e77. [CrossRef] [PubMed]
31. Lindskog, E.K.; Fischer, S.; Wenger, T.; Schulz, P. Chapter 6—Host cells. In *Biopharmaceutical Processing*; Jagschies, G., Lindskog, E., Łącki, K., Galliher, P., Eds.; Elsevier: Amsterdam, The Netherlands, 2018; pp. 111–130.
32. Fliedl, L.; Grillari, J.; Grillari-Voglauer, R. Human cell lines for the production of recombinant proteins: On the horizon. *New Biotechnol.* **2015**, *32*, 673–679. [CrossRef]
33. Matsuda, T.; Ito, T.; Takemoto, C.; Katsura, K.; Ikeda, M.; Wakiyama, M.; Kukimoto-Niino, M.; Yokoyama, S.; Kurosawa, Y.; Shirouzu, M. Cell-free synthesis of functional antibody fragments to provide a structural basis for antibody–antigen interaction. *PLoS ONE* **2018**, *13*, e0193158. [CrossRef]
34. Stech, M.; Kubick, S. Cell-free synthesis meets antibody production: A review. *Antibodies* **2015**, *4*, 12. [CrossRef]

35. Pujols, J.; Peña-Díaz, S.; Ventura, S. Aggrescan3d: Toward the prediction of the aggregation propensities of protein structures. In *Methods Mol. Biol.*; Humana Press Inc.: New York, NY, USA, 2018; Volume 1762, pp. 427–443.
36. Van der Kant, R.; Karow-Zwick, A.R.; Van Durme, J.; Blech, M.; Gallardo, R.; Seeliger, D.; Aßfalg, K.; Baatsen, P.; Compernolle, G.; Gils, A.; et al. Prediction and reduction of the aggregation of monoclonal antibodies. *J. Mol. Biol.* **2017**, *429*, 1244–1261. [CrossRef]
37. Lerch, T.F.; Sharpe, P.; Mayclin, S.J.; Edwards, T.E.; Lee, E.; Conlon, H.D.; Polleck, S.; Rouse, J.C.; Luo, Y.; Zou, Q. Infliximab crystal structures reveal insights into self-association. *mAbs* **2017**, *9*, 874–883. [CrossRef]
38. Dobson, C.L.; Devine, P.W.A.; Phillips, J.J.; Higazi, D.R.; Lloyd, C.; Popovic, B.; Arnold, J.; Buchanan, A.; Lewis, A.; Goodman, J.; et al. Engineering the surface properties of a human monoclonal antibody prevents self-association and rapid clearance in vivo. *Sci. Rep.* **2016**, *6*, 38644. [CrossRef]
39. Courtois, F.; Agrawal, N.J.; Lauer, T.M.; Trout, B.L. Rational design of therapeutic mabs against aggregation through protein engineering and incorporation of glycosylation motifs applied to bevacizumab. *mAbs* **2016**, *8*, 99–112. [CrossRef]
40. Courtois, F.; Schneider, C.P.; Agrawal, N.J.; Trout, B.L. Rational design of biobetters with enhanced stability. *J. Pharm. Sci.* **2015**, *104*, 2433–2440. [CrossRef]
41. Yu, M.; Wu, J.; Shi, J.; Farokhzad, O.C. Nanotechnology for protein delivery: Overview and perspectives. *J. Control. Release* **2016**, *240*, 24–37. [CrossRef]
42. Shi, J.; Kantoff, P.W.; Wooster, R.; Farokhzad, O.C. Cancer nanomedicine: Progress, challenges and opportunities. *Nat. Rev. Cancer* **2016**, *17*, 20. [CrossRef]
43. Dalziel, M.; Beers, S.A.; Cragg, M.S.; Crispin, M. Through the barricades: Overcoming the barriers to effective antibody-based cancer therapeutics. *Glycobiology* **2018**, *28*, 697–712. [CrossRef]
44. Lambert, J.M.; Berkenblit, A. Antibody-drug conjugates for cancer treatment. *Annu. Rev. Med.* **2018**, *69*, 191–207. [CrossRef]
45. Lambert, J.M.; Morris, C.Q. Antibody–drug conjugates (adcs) for personalized treatment of solid tumors: A review. *Adv. Ther.* **2017**, *34*, 1015–1035. [CrossRef]
46. Levin, D.; Golding, B.; Strome, S.E.; Sauna, Z.E. Fc fusion as a platform technology: Potential for modulating immunogenicity. *Trends Biotechnol.* **2015**, *33*, 27–34. [CrossRef]
47. Henry, K.A. Next-generation DNA sequencing of vh/vl repertoires: A primer and guide to applications in single-domain antibody discovery. In *Phage Display: Methods and Protocols*; Hust, M., Lim, T.S., Eds.; Springer: New York, NY, USA, 2018; pp. 425–446.
48. Frenzel, A.; Roskos, L.; Klakamp, S.; Liang, M.; Arends, R.; Green, L. Antibody affinity. In *Handbook of Therapeutic Antibodies*, 2nd ed.; Wiley Blackwell: Hoboken, NJ, USA, 2014; Volume 1–4, pp. 115–140.
49. Hu, J.; Han, J.; Li, H.; Zhang, X.; Liu, L.; Chen, F.; Zeng, B. Human embryonic kidney 293 cells: A vehicle for biopharmaceutical manufacturing, structural biology, and electrophysiology. *Cells Tissues Organs* **2018**, *205*, 1–8. [CrossRef]
50. Jacobi, A.; Enenkel, B.; Garidel, P.; Eckermann, C.; Knappenberger, M.; Presser, I.; Kaufmann, H. Process development and manufacturing of therapeutic antibodies. In *Handbook of Therapeutic Antibodies*, 2nd ed.; Wiley Blackwell: Hoboken, NJ, USA, 2014; Volume 2–4, pp. 601–664.
51. Kayser, V.; Chennamsetty, N.; Voynov, V.; Helk, B.; Forrer, K.; Trout, B.L. A screening tool for therapeutic monoclonal antibodies: Identifying the most stable protein and its best formulation based on thioflavin t binding. *Biotechnol. J.* **2012**, *7*, 127–132. [CrossRef]
52. Tada, M.; Tatematsu, K.-I.; Ishii-Watabe, A.; Harazono, A.; Takakura, D.; Hashii, N.; Sezutsu, H.; Kawasaki, N. Characterization of anti-cd20 monoclonal antibody produced by transgenic silkworms (bombyx mori). *mAbs* **2015**, *7*, 1138–1150. [CrossRef]
53. Maccani, A.; Landes, N.; Stadlmayr, G.; Maresch, D.; Leitner, C.; Maurer, M.; Gasser, B.; Ernst, W.; Kunert, R.; Mattanovich, D. Pichia pastoris secretes recombinant proteins less efficiently than chinese hamster ovary cells but allows higher space-time yields for less complex proteins. *Biotechnol. J.* **2014**, *9*, 526–537. [CrossRef]
54. Frenzel, A.; Hust, M.; Schirrmann, T. Expression of recombinant antibodies. *Front. Immunol.* **2013**, *4*, 217. [CrossRef]
55. Thoring, L.; Kubick, S. Versatile cell-free protein synthesis systems based on chinese hamster ovary cells. In *Recombinant Protein Expression in Mammalian Cells: Methods and Protocols*; Hacker, D.L., Ed.; Springer: New York, NY, USA, 2018; pp. 289–308.

56. Tran, K.; Gurramkonda, C.; Cooper, M.A.; Pilli, M.; Taris, J.E.; Selock, N.; Han, T.-C.; Tolosa, M.; Zuber, A.; Peñalber-Johnstone, C.; et al. Cell-free production of a therapeutic protein: Expression, purification, and characterization of recombinant streptokinase using a cho lysate. *Biotechnol. Bioeng.* **2018**, *115*, 92–102. [CrossRef]
57. Stech, M.; Nikolaeva, O.; Thoring, L.; Stöcklein, W.F.M.; Wüstenhagen, D.A.; Hust, M.; Dübel, S.; Kubick, S. Cell-free synthesis of functional antibodies using a coupled in vitro transcription-translation system based on cho cell lysates. *Sci. Rep.* **2017**, *7*, 12030. [CrossRef]
58. Li, J.; Wang, H.; Kwon, Y.-C.; Jewett, M.C. Establishing a high yielding streptomyces-based cell-free protein synthesis system. *Biotechnol. Bioeng.* **2017**, *114*, 1343–1353. [CrossRef]
59. Jaroentomeechai, T.; Stark, J.C.; Natarajan, A.; Glasscock, C.J.; Yates, L.E.; Hsu, K.J.; Mrksich, M.; Jewett, M.C.; DeLisa, M.P. Single-pot glycoprotein biosynthesis using a cell-free transcription-translation system enriched with glycosylation machinery. *Nat. Commun.* **2018**, *9*, 2686. [CrossRef]
60. Gurramkonda, C.; Rao, A.; Borhani, S.; Pilli, M.; Deldari, S.; Ge, X.; Pezeshk, N.; Han, T.-C.; Tolosa, M.; Kostov, Y.; et al. Improving the recombinant human erythropoietin glycosylation using microsome supplementation in cho cell-free system. *Biotechnol. Bioeng.* **2018**, *115*, 1253–1264. [CrossRef]
61. Nakamura, T.; Omasa, T. Optimization of cell line development in the gs-cho expression system using a high-throughput, single cell-based clone selection system. *J. Biosci. Bioeng.* **2015**, *120*, 323–329. [CrossRef]
62. Zhou, Y.; Raju, R.; Alves, C.; Gilbert, A. Debottlenecking protein secretion and reducing protein aggregation in the cellular host. *Curr. Opin. Biotechnol.* **2018**, *53*, 151–157. [CrossRef]
63. Inwood, S.; Betenbaugh, M.J.; Lal, M.; Shiloach, J. Genome-wide high-throughput rnai screening for identification of genes involved in protein production. In *Recombinant Protein Expression in Mammalian Cells: Methods and Protocols*; Hacker, D.L., Ed.; Springer: New York, NY, USA, 2018; pp. 209–219.
64. Dangi, A.K.; Sinha, R.; Dwivedi, S.; Gupta, S.K.; Shukla, P. Cell line techniques and gene editing tools for antibody production: A review. *Front. Pharmacol.* **2018**, *9*, 630. [CrossRef]
65. Delic, M.; Göngrich, R.; Mattanovich, D.; Gasser, B. Engineering of protein folding and secretion—Strategies to overcome bottlenecks for efficient production of recombinant proteins. *Antioxid. Redox Signal.* **2014**, *21*, 414–437. [CrossRef]
66. Zhong, X.; Wright, J.F. Biological insights into therapeutic protein modifications throughout trafficking and their biopharmaceutical applications. *Int. J. Cell Biol.* **2013**, *2013*, 19. [CrossRef]
67. Altamirano, C.; Berrios, J.; Vergara, M.; Becerra, S. Advances in improving mammalian cells metabolism for recombinant protein production. *Electron. J. Biotechnol.* **2013**, *16*. [CrossRef]
68. Parola, C.; Mason, D.M.; Zingg, A.; Neumeier, D.; Reddy, S.T. Genome engineering of hybridomas to generate stable cell lines for antibody expression. In *Recombinant Protein Expression in Mammalian Cells: Methods and Protocols*; Hacker, D.L., Ed.; Springer: New York, NY, USA, 2018; pp. 79–111.
69. You, M.; Yang, Y.; Zhong, C.; Chen, F.; Wang, X.; Jia, T.; Chen, Y.; Zhou, B.; Mi, Q.; Zhao, Q.; et al. Efficient mab production in cho cells with optimized signal peptide, codon, and utr. *Appl. Microbiol. Biotechnol.* **2018**, *102*, 5953–5964. [CrossRef]
70. Mauro, V.P.; Chappell, S.A. Considerations in the use of codon optimization for recombinant protein expression. In *Recombinant Protein Expression in Mammalian Cells: Methods and Protocols*; Hacker, D.L., Ed.; Springer: New York, NY, USA, 2018; pp. 275–288.
71. Michael, I.P.; Nagy, A. Inducible protein production in 293 cells using the piggybac transposon system. In *Recombinant Protein Expression in Mammalian Cells: Methods and Protocols*; Hacker, D.L., Ed.; Springer: New York, NY, USA, 2018; pp. 57–68.
72. Balasubramanian, S. Recombinant cho cell pool generation using piggybac transposon system. In *Recombinant Protein Expression in Mammalian Cells: Methods and Protocols*; Hacker, D.L., Ed.; Springer: New York, NY, USA, 2018; pp. 69–78.
73. Jäger, V.; Büssow, K.; Schirrmann, T. Transient recombinant protein expression in mammalian cells. In *Animal Cell Culture*; Al-Rubeai, M., Ed.; Springer International Publishing: Cham, Switzerland, 2015; pp. 27–64.
74. Wijesuriya, S.D.; Pongo, E.; Tomic, M.; Zhang, F.; Garcia-Rodriquez, C.; Conrad, F.; Farr-Jones, S.; Marks, J.D.; Horwitz, A.H. Antibody engineering to improve manufacturability. *Protein Expression Purif.* **2018**, *149*, 75–83. [CrossRef]

75. L'Abbé, D.; Bisson, L.; Gervais, C.; Grazzini, E.; Durocher, Y. Transient gene expression in suspension hek293-ebna1 cells. In *Recombinant Protein Expression in Mammalian Cells: Methods and Protocols*; Hacker, D.L., Ed.; Springer: New York, NY, USA, 2018; pp. 1–16.
76. Arena, T.A.; Harms, P.D.; Wong, A.W. High throughput transfection of hek293 cells for transient protein production. In *Recombinant Protein Expression in Mammalian Cells: Methods and Protocols*; Hacker, D.L., Ed.; Springer: New York, NY, USA, 2018; pp. 179–187.
77. Agrawal, V.; Yu, B.; Pagila, R.; Yang, B.; Simonsen, C.; Beske, O. A high-yielding, cho-k1-based transient transfection system. *BioProcess Int.* **2013**, *11*, 28–35.
78. Hacker, D.L.; Kiseljak, D.; Rajendra, Y.; Thurnheer, S.; Baldi, L.; Wurm, F.M. Polyethyleneimine-based transient gene expression processes for suspension-adapted hek-293e and cho-dg44 cells. *Protein Expression Purif.* **2013**, *92*, 67–76. [CrossRef]
79. Ritacco, F.V.; Wu, Y.; Khetan, A. Cell culture media for recombinant protein expression in chinese hamster ovary (cho) cells: History, key components, and optimization strategies. *Biotechnol. Prog.* **2018**, *34*, 1407–1426. [CrossRef]
80. Whitford, W.G.; Lundgren, M.; Fairbank, A. Chapter 8—Cell culture media in bioprocessing. In *Biopharmaceutical Processing*; Jagschies, G., Lindskog, E., Łącki, K., Galliher, P., Eds.; Elsevier: Amsterdam, The Netherlands, 2018; pp. 147–162.
81. Davami, F.; Eghbalpour, F.; Barkhordari, F.; Mahboudi, F. Effect of peptone feeding on transient gene expression process in cho dg44. *Avicenna J. Med. Biotechnol.* **2014**, *6*, 147–155.
82. Mahboudi, F.; Abolhassan, M.R.; Azarpanah, A.; Aghajani-Lazarjani, H.; Sadeghi-Haskoo, M.A.; Maleknia, S.; Vaziri, B. The role of different supplements in expression level of monoclonal antibody against human cd20. *Avicenna J. Med. Biotechnol.* **2013**, *5*, 140–147.
83. You, M.; Liu, Y.; Chen, Y.; Guo, J.; Wu, J.; Fu, Y.; Shen, R.; Qi, R.; Luo, W.; Xia, N. Maximizing antibody production in suspension-cultured mammalian cells by the customized transient gene expression method. *Biosci. Biotechnol. Biochem.* **2013**, *77*, 1207–1213. [CrossRef]
84. Backliwal, G.; Hildinger, M.; Kuettel, I.; Delegrange, F.; Hacker, D.L.; Wurm, F.M. Valproic acid: A viable alternative to sodium butyrate for enhancing protein expression in mammalian cell cultures. *Biotechnol. Bioeng.* **2008**, *101*, 182–189. [CrossRef]
85. Elgundi, Z.; Sifniotis, V.; Reslan, M.; Cruz, E.; Kayser, V. Laboratory scale production and purification of a therapeutic antibody. *JoVE* **2017**. [CrossRef]
86. Kim, S.J.; Ha, G.S.; Lee, G.; Lim, S.I.; Lee, C.M.; Yang, Y.H.; Lee, J.; Kim, J.E.; Lee, J.H.; Shin, Y.; et al. Enhanced expression of soluble antibody fragments by low-temperature and overdosing with a nitrogen source. *Enzyme Microb. Technol.* **2018**, *115*, 9–15. [CrossRef]
87. Lalonde, M.-E.; Durocher, Y. Therapeutic glycoprotein production in mammalian cells. *J. Biotechnol.* **2017**, *251*, 128–140. [CrossRef]
88. Rajendra, Y.; Kiseljak, D.; Baldi, L.; Hacker, D.L.; Wurm, F.M. A simple high-yielding process for transient gene expression in cho cells. *J. Biotechnol.* **2011**, *153*, 22–26. [CrossRef]
89. Codamo, J.; Munro, T.P.; Hughes, B.S.; Song, M.; Gray, P.P. Enhanced cho cell-based transient gene expression with the epi-cho expression system. *Mol. Biotechnol.* **2011**, *48*, 109–115. [CrossRef]
90. Codamo, J.; Hou, J.J.C.; Hughes, B.S.; Gray, P.P.; Munro, T.P. Efficient mab production in cho cells incorporating pei-mediated transfection, mild hypothermia and the co-expression of xbp-1. *J. Chem. Technol. Biotechnol.* **2011**, *86*, 923–934. [CrossRef]
91. Castan, A.; Schulz, P.; Wenger, T.; Fischer, S. Chapter 7—Cell line development. In *Biopharmaceutical Processing*; Jagschies, G., Lindskog, E., Łącki, K., Galliher, P., Eds.; Elsevier: Amsterdam, The Netherlands, 2018; pp. 131–146.
92. Lindskog, E.K. Chapter 31—The upstream process: Principal modes of operation. In *Biopharmaceutical Processing*; Jagschies, G., Lindskog, E., Łącki, K., Galliher, P., Eds.; Elsevier: Amsterdam, The Netherlands, 2018; pp. 625–635.
93. Fisher, A.C.; Kamga, M.-H.; Agarabi, C.; Brorson, K.; Lee, S.L.; Yoon, S. The current scientific and regulatory landscape in advancing integrated continuous biopharmaceutical manufacturing. *Trends Biotechnol.* **2018**, *37*, 253–267. [CrossRef]
94. Rathore, A.S.; Agarwal, H.; Sharma, A.K.; Pathak, M.; Muthukumar, S. Continuous processing for production of biopharmaceuticals. *Prep. Biochem. Biotechnol.* **2015**, *45*, 836–849. [CrossRef]

95. Shukla, A.A.; Suda, E. Chapter 3—Harvest and recovery of monoclonal antibodies: Cell removal and clarification. In *Process Scale Purification of Antibodies*; Gottschalk, U., Ed.; John Wiley & Sons, Inc.: Hoboken, NJ, USA, 2017; pp. 55–79.
96. Thömmes, J.; Twyman, R.M.; Gottschalk, U. Chapter 10—Alternatives to packed-bed chromatography for antibody extraction and purification. In *Process Scale Purification of Antibodies*; Gottschalk, U., Ed.; John Wiley & Sons: Hoboken, NJ, USA, 2017; pp. 215–231.
97. Glynn, J. Chapter 11—Process-scale precipitation of impurities in mammalian cell culture broth. In *Process Scale Purification of Antibodies*; Gottschalk, U., Ed.; John Wiley & Sons, Inc.: Hoboken, NJ, USA, 2017; pp. 233–246.
98. Singh, N.; Arunkumar, A.; Chollangi, S.; Tan, Z.G.; Borys, M.; Li, Z.J. Clarification technologies for monoclonal antibody manufacturing processes: Current state and future perspectives. *Biotechnol. Bioeng.* **2016**, *113*, 698–716. [CrossRef]
99. Singh, N.; Chollangi, S. Chapter 4—Next-generation clarification technologies for the downstream processing of antibodies. In *Process scale Purification of Antibodies*; Gottschalk, U., Ed.; John Wiley & Sons, Inc.: Hoboken, NJ, USA, 2017; pp. 81–112.
100. Kelley, B. Chapter 1—Downstream processing of monoclonal antibodies: Current practices and future opportunities. In *Process Scale Purification of Antibodies*; Gottschalk, U., Ed.; John Wiley & Sons, Inc.: Hoboken, NJ, USA, 2017; pp. 1–21.
101. Danielsson, Å. Chapter 17—Affinity chromatography. In *Biopharmaceutical Processing*; Jagschies, G., Lindskog, E., Łącki, K., Galliher, P., Eds.; Elsevier: Amsterdam, The Netherlands, 2018; pp. 367–378.
102. Kinna, A.; Tolner, B.; Rota, E.M.; Titchener-Hooker, N.; Nesbeth, D.; Chester, K. Imac capture of recombinant protein from unclarified mammalian cell feed streams. *Biotechnol. Bioeng.* **2016**, *113*, 130–140. [CrossRef]
103. Ghose, S.; Jin, M.; Liu, J.; Hickey, J.; Lee, S. Chapter 14—Integrated polishing steps for monoclonal antibody purification. In *Process Scale Purification of Antibodies*; Gottschalk, U., Ed.; John Wiley & Sons, Inc.: Hoboken, NJ, USA, 2017; pp. 303–323.
104. Joshi, V.; Shivach, T.; Kumar, V.; Yadav, N.; Rathore, A. Avoiding antibody aggregation during processing: Establishing hold times. *Biotechnol. J.* **2014**, *9*, 1195–1205. [CrossRef]
105. Liderfelt, J.; Royce, J. Chapter 23—Filtration methods for use in purification processes (concentration and buffer exchange). In *Biopharmaceutical Processing*; Jagschies, G., Lindskog, E., Łącki, K., Galliher, P., Eds.; Elsevier: Amsterdam, The Netherlands, 2018; pp. 441–453.
106. Anselmo, A.C.; Gokarn, Y.; Mitragotri, S. Non-invasive delivery strategies for biologics. *Nat. Rev. Drug Discov.* **2018**, *18*, 19. [CrossRef]
107. Sousa, D.; Ferreira, D.; Rodrigues, J.L.; Rodrigues, L.R. Chapter 14—Nanotechnology in targeted drug delivery and therapeutics. In *Applications of Targeted Nano Drugs and Delivery Systems*; Mohapatra, S.S., Ranjan, S., Dasgupta, N., Mishra, R.K., Thomas, S., Eds.; Elsevier: Amsterdam, The Netherlands, 2019; pp. 357–409.
108. Jani, R.; Krupa, G.; Rupal, J. Active targeting of nanoparticles: An innovative technology for drug delivery in cancer therapeutics. *J. Drug Deliv. Ther.* **2019**. [CrossRef]
109. Abdelaziz, H.M.; Gaber, M.; Abd-Elwakil, M.M.; Mabrouk, M.T.; Elgohary, M.M.; Kamel, N.M.; Kabary, D.M.; Freag, M.S.; Samaha, M.W.; Mortada, S.M.; et al. Inhalable particulate drug delivery systems for lung cancer therapy: Nanoparticles, microparticles, nanocomposites and nanoaggregates. *J. Control. Release* **2018**, *269*, 374–392. [CrossRef]
110. Jackisch, C.; Kim, S.B.; Semiglazov, V.; Melichar, B.; Pivot, X.; Hillenbach, C.; Stroyakovskiy, D.; Lum, B.L.; Elliott, R.; Weber, H.A.; et al. Subcutaneous versus intravenous formulation of trastuzumab for her2-positive early breast cancer: Updated results from the phase iii hannah study. *Ann. Oncol.* **2015**, *26*, 320–325. [CrossRef]
111. Lambertini, M.; Pondé, N.F.; Solinas, C.; de Azambuja, E. Adjuvant trastuzumab: A 10-year overview of its benefit. *Expert Rev. Anticancer Ther.* **2017**, *17*, 61–74. [CrossRef]
112. Sanford, M. Subcutaneous trastuzumab: A review of its use in her2-positive breast cancer. *Target. Oncol.* **2014**, *9*, 85–94. [CrossRef]
113. Reichert, J.M. Adalimumab (humira®). In *Handbook of Therapeutic Antibodies*, 2nd ed.; Wiley Blackwell: Hoboken, NJ, USA, 2014; Volume 3–4, pp. 1309–1322.

114. Crommelin, D.J.A.; Hawe, A.; Jiskoot, W. Formulation of biologics including biopharmaceutical considerations. In *Pharmaceutical Biotechnology: Fundamentals and Applications*; Crommelin, D.J.A., Sindelar, R.D., Meibohm, B., Eds.; Springer International Publishing: Cham, Switzerland, 2019; pp. 83–103.
115. Singh, S.K.; Mahler, H.-C.; Hartman, C.; Stark, C.A. Are injection site reactions in monoclonal antibody therapies caused by polysorbate excipient degradants? *J. Pharm. Sci.* **2018**, *107*, 2735–2741. [CrossRef]
116. Rayaprolu, B.M.; Strawser, J.J.; Anyarambhatla, G. Excipients in parenteral formulations: Selection considerations and effective utilization with small molecules and biologics. *Drug Dev. Ind. Pharm.* **2018**, *44*, 1565–1571. [CrossRef]
117. Pimentel, F.F.; Morgan, G.; Tiezzi, D.G.; de Andrade, J.M. Development of new formulations of biologics: Expectations, immunogenicity, and safety for subcutaneous trastuzumab. *Pharmaceut. Med.* **2018**, *32*, 319–325. [CrossRef]
118. Garidel, P.; Kuhn, A.B.; Schäfer, L.V.; Karow-Zwick, A.R.; Blech, M. High-concentration protein formulations: How high is high? *Eur. J. Pharm. Biopharm.* **2017**, *119*, 353–360. [CrossRef]
119. Kemter, K.; Altrichter, J.; Derwand, R.; Kriehuber, T.; Reinauer, E.; Scholz, M. Amino acid-based advanced liquid formulation development for highly concentrated therapeutic antibodies balances physical and chemical stability and low viscosity. *Biotechnol. J.* **2018**, *13*, e1700523. [CrossRef]
120. Mandal, A.; Pal, D.; Agrahari, V.; Trinh, H.M.; Joseph, M.; Mitra, A.K. Ocular delivery of proteins and peptides: Challenges and novel formulation approaches. *Adv. Drug Del. Rev.* **2018**, *126*, 67–95. [CrossRef]
121. Reslan, M.; Kayser, V. The effect of deuterium oxide on the conformational stability and aggregation of bovine serum albumin. *Pharm. Dev. Technol.* **2016**, 1–7. [CrossRef]
122. Reslan, M.; Ranganathan, V.; Macfarlane, D.R.; Kayser, V. Choline ionic liquid enhances the stability of herceptin(r) (trastuzumab). *Chem. Commun. (Camb.)* **2018**, *54*, 10622–10625. [CrossRef]
123. Reslan, M.; Kayser, V. Ionic liquids as biocompatible stabilizers of proteins. *Biophys. Rev.* **2018**, *10*, 781–793. [CrossRef]
124. Emami, F.; Vatanara, A.; Park, E.J.; Na, D.H. Drying technologies for the stability and bioavailability of biopharmaceuticals. *Pharmaceutics* **2018**, *10*, 131. [CrossRef]
125. Izutsu, K.I. Applications of freezing and freeze-drying in pharmaceutical formulations. In *Adv. Exp. Med. Biol.*; Springer: New York, NY, USA, 2018; Volume 1081, pp. 371–383.
126. Schermeyer, M.T.; Wöll, A.K.; Kokke, B.; Eppink, M.; Hubbuch, J. Characterization of highly concentrated antibody solution—A toolbox for the description of protein long-term solution stability. *mAbs* **2017**, *9*, 1169–1185. [CrossRef]
127. Kayser, V.; Chennamsetty, N.; Voynov, V.; Helk, B.; Forrer, K.; Trout, B.L. Evaluation of a non-arrhenius model for therapeutic monoclonal antibody aggregation. *J. Pharm. Sci.* **2011**, *100*, 2526–2542. [CrossRef]
128. Rabia, L.A.; Desai, A.A.; Jhajj, H.S.; Tessier, P.M. Understanding and overcoming trade-offs between antibody affinity, specificity, stability and solubility. *Biochem. Eng. J.* **2018**, *137*, 365–374. [CrossRef]
129. Van de Bovenkamp, F.S.; Derksen, N.I.L.; van Breemen, M.J.; de Taeye, S.W.; Ooijevaar-de Heer, P.; Sanders, R.W.; Rispens, T. Variable domain n-linked glycans acquired during antigen-specific immune responses can contribute to immunoglobulin g antibody stability. *Front. Immunol.* **2018**, *9*, 740. [CrossRef]
130. Kuhn, A.B.; Kube, S.; Karow-Zwick, A.R.; Seeliger, D.; Garidel, P.; Blech, M.; Schäfer, L.V. Improved solution-state properties of monoclonal antibodies by targeted mutations. *J. Phys. Chem. B* **2017**, *121*, 10818–10827. [CrossRef]
131. Singh, S.N.; Yadav, S.; Shire, S.J.; Kalonia, D.S. Dipole-dipole interaction in antibody solutions: Correlation with viscosity behavior at high concentration. *Pharm. Res.* **2014**, *31*, 2549–2558. [CrossRef]
132. Ionescu, R.M.; Vlasak, J.; Price, C.; Kirchmeier, M. Contribution of variable domains to the stability of humanized igg1 monoclonal antibodies. *J. Pharm. Sci.* **2008**, *97*, 1414–1426. [CrossRef]
133. Liu, L. Pharmacokinetics of monoclonal antibodies and fc-fusion proteins. *Protein Cell* **2018**, *9*, 15–32. [CrossRef]
134. Schweizer, D.; Serno, T.; Goepferich, A. Controlled release of therapeutic antibody formats. *Eur. J. Pharm. Biopharm.* **2014**, *88*, 291–309. [CrossRef]
135. Rudnick, S.I.; Adams, G.P. Affinity and avidity in antibody-based tumor targeting. *Cancer Biother. Radiopharm.* **2009**, *24*, 155–161. [CrossRef]
136. Ickenstein, L.M.; Garidel, P. Hydrogel formulations for biologicals: Current spotlight from a commercial perspective. *Ther. Deliv.* **2018**, *9*, 221–230. [CrossRef]

137. Saxena, A.; Wu, D. Advances in therapeutic fc engineering—Modulation of igg-associated effector functions and serum half-life. *Front. Immunol.* **2016**, *7*, 580. [CrossRef]
138. Jefferis, R. Glycosylation of antibody molecules. In *Handbook of Therapeutic Antibodies*, 2nd ed.; Wiley Blackwell: Hoboken, NJ, USA, 2014; Volume 1–4, pp. 171–200.
139. Zheng, K.; Bantog, C.; Bayer, R. The impact of glycosylation on monoclonal antibody conformation and stability. *mAbs* **2011**, *3*, 568–576. [CrossRef]
140. Lei, C.; Gong, R.; Ying, T. Editorial: Antibody fc engineering: Towards better therapeutics. *Front. Immunol.* **2018**, *9*, 2450. [CrossRef]
141. Booth, B.J.; Ramakrishnan, B.; Narayan, K.; Wollacott, A.M.; Babcock, G.J.; Shriver, Z.; Viswanathan, K. Extending human igg half-life using structure-guided design. *mAbs* **2018**, *10*, 1098–1110. [CrossRef]
142. Kellner, C.; Otte, A.; Cappuzzello, E.; Klausz, K.; Peipp, M. Modulating cytotoxic effector functions by fc engineering to improve cancer therapy. *Transfus. Med. Hemoth.* **2017**, *44*, 327–336. [CrossRef]
143. Wang, Q.; Chen, Y.; Pelletier, M.; Cvitkovic, R.; Bonnell, J.; Chang, C.Y.; Koksal, A.C.; O'Connor, E.; Gao, X.; Yu, X.Q.; et al. Enhancement of antibody functions through fc multiplications. *mAbs* **2017**, *9*, 393–403. [CrossRef]
144. Lefranc, M.P.; Ehrenmann, F.; Kossida, S.; Giudicelli, V.; Duroux, P. Use of imgt® databases and tools for antibody engineering and humanization. In *Methods Mol. Biol.*; Humana Press Inc.: New York, NY, USA, 2018; Volume 1827, pp. 35–69.
145. Lefranc, M.-P.; Giudicelli, V.; Duroux, P.; Jabado-Michaloud, J.; Folch, G.; Aouinti, S.; Carillon, E.; Duvergey, H.; Houles, A.; Paysan-Lafosse, T.; et al. Imgt(®), the international immunogenetics information system(®) 25 years on. *Nucleic Acids Res.* **2015**, *43*, D413–D422. [CrossRef]
146. Martin, A.C.R.; Allen, J. Bioinformatics tools for analysis of antibodies. In *Handbook of Therapeutic Antibodies*, 2nd ed.; Wiley Blackwell: Hoboken, NJ, USA, 2014; Volume 1–4, pp. 201–228.
147. Śledź, P.; Caflisch, A. Protein structure-based drug design: From docking to molecular dynamics. *Curr. Opin. Struct. Biol.* **2018**, *48*, 93–102. [CrossRef]
148. Pandya, A.; Howard, M.J.; Zloh, M.; Dalby, P.A. An evaluation of the potential of nmr spectroscopy and computational modelling methods to inform biopharmaceutical formulations. *Pharmaceutics* **2018**, *10*, 1–24. [CrossRef]
149. Rathore, D.; Faustino, A.; Schiel, J.; Pang, E.; Boyne, M.; Rogstad, S. The role of mass spectrometry in the characterization of biologic protein products. *Expert Rev. Proteomic.* **2018**, *15*, 431–449. [CrossRef]
150. Wang, X.; An, Z.; Luo, W.; Xia, N.; Zhao, Q. Molecular and functional analysis of monoclonal antibodies in support of biologics development. *Protein Cell* **2018**, *9*, 74–85. [CrossRef]
151. Baker Edward, N. Crystallography and the development of therapeutic medicines. *IUCrJ* **2018**, *5*, 118–119. [CrossRef]
152. Vénien-Bryan, C.; Li, Z.; Vuillard, L.; Boutin, J.A. Cryo-electron microscopy and X-ray crystallography: Complementary approaches to structural biology and drug discovery. *Acta Crystallogr. Sect. F Struct. Biol. Commun.* **2017**, *73*, 174–183. [CrossRef]
153. Brader, M.L.; Baker, E.N.; Dunn, M.F.; Laue, T.M.; Carpenter, J.F. Using x-ray crystallography to simplify and accelerate biologics drug development. *J. Pharm. Sci.* **2017**, *106*, 477–494. [CrossRef]
154. Nero, T.L.; Parker, M.W.; Morton, C.J. Protein structure and computational drug discovery. *Biochem. Soc. Trans.* **2018**, *46*, 1367–1379. [CrossRef]
155. Blaffert, J.; Haeri, H.H.; Blech, M.; Hinderberger, D.; Garidel, P. Spectroscopic methods for assessing the molecular origins of macroscopic solution properties of highly concentrated liquid protein solutions. *Anal. Biochem.* **2018**, *561–562*, 70–88. [CrossRef]
156. Brinson, R.G.; Marino, J.P.; Delaglio, F.; Arbogast, L.W.; Evans, R.M.; Kearsley, A.; Gingras, G.; Ghasriani, H.; Aubin, Y.; Pierens, G.K.; et al. Enabling adoption of 2d-nmr for the higher order structure assessment of monoclonal antibody therapeutics. *mAbs* **2019**, *11*, 94–105. [CrossRef]
157. Young, J.A.; Gabrielson, J.P. Higher order structure methods for similarity assessment. In *Biosimilars: Regulatory, Clinical, and Biopharmaceutical Development*; Gutka, H.J., Yang, H., Kakar, S., Eds.; Springer International Publishing: Cham, Switzerland, 2018; pp. 321–337.
158. Kumar, S.; Plotnikov, N.V.; Rouse, J.C.; Singh, S.K. Biopharmaceutical informatics: Supporting biologic drug development via molecular modelling and informatics. *J. Pharm. Pharmacol.* **2018**, *70*, 595–608. [CrossRef]

159. Westbrook, J.D.; Burley, S.K. How structural biologists and the protein data bank contributed to recent fda new drug approvals. *Structure* **2018**, *27*, 211–217. [CrossRef]
160. Burley, S.K.; Berman, H.M.; Christie, C.; Duarte, J.M.; Feng, Z.; Westbrook, J.; Young, J.; Zardecki, C. Rcsb protein data bank: Sustaining a living digital data resource that enables breakthroughs in scientific research and biomedical education. *Protein Sci.* **2018**, *27*, 316–330. [CrossRef]
161. Kuroda, D.; Tsumoto, K. Antibody affinity maturation by computational design. In *Methods Mol. Biol.*; Humana Press Inc.: New York, NY, USA, 2018; Volume 1827, pp. 15–34.
162. Fischman, S.; Ofran, Y. Computational design of antibodies. *Curr. Opin. Struct. Biol.* **2018**, *51*, 156–162. [CrossRef]
163. Adolf-Bryfogle, J.; Kalyuzhniy, O.; Kubitz, M.; Weitzner, B.D.; Hu, X.; Adachi, Y.; Schief, W.R.; Dunbrack, R.L., Jr. Rosettaantibodydesign (rabd): A general framework for computational antibody design. *PLoS Comp. Biol.* **2018**, *14*, e1006112. [CrossRef]
164. Branco, R.J.; Dias, A.M.; Roque, A.C. Understanding the molecular recognition between antibody fragments and protein a biomimetic ligand. *J. Chromatogr. A* **2012**, *1244*, 106–115. [CrossRef]
165. Huang, B.; Liu, F.F.; Dong, X.Y.; Sun, Y. Molecular mechanism of the affinity interactions between protein a and human immunoglobulin g1 revealed by molecular simulations. *J. Phys. Chem. B* **2011**, *115*, 4168–4176. [CrossRef]
166. Van der Kant, R.; van Durme, J.; Rousseau, F.; Schymkowitz, J. Solubis: Optimizing protein solubility by minimal point mutations. In *Methods Mol. Biol.*; Humana Press Inc.: New York, NY, USA, 2019; Volume 1873, pp. 317–333.
167. Gil-Garcia, M.; Banó-Polo, M.; Varejao, N.; Jamroz, M.; Kuriata, A.; Díaz-Caballero, M.; Lascorz, J.; Morel, B.; Navarro, S.; Reverter, D.; et al. Combining structural aggregation propensity and stability predictions to redesign protein solubility. *Mol. Pharm.* **2018**, *15*, 3846–3859. [CrossRef]
168. Seeliger, D.; Schulz, P.; Litzenburger, T.; Spitz, J.; Hoerer, S.; Blech, M.; Enenkel, B.; Studts, J.M.; Garidel, P.; Karow, A.R. Boosting antibody developability through rational sequence optimization. *mAbs* **2015**, *7*, 505–515. [CrossRef]
169. Chennamsetty, N.; Voynov, V.; Kayser, V.; Helk, B.; Trout, B.L. Prediction of aggregation prone regions of therapeutic proteins. *J. Phys. Chem. B* **2010**, *114*, 6614–6624. [CrossRef]
170. Chennamsetty, N.; Voynov, V.; Kayser, V.; Helk, B.; Trout, B.L. Design of therapeutic proteins with enhanced stability. *Proc. Natl. Acad. Sci. USA* **2009**, *106*, 11937–11942. [CrossRef]
171. Majumder, S.; Jones, M.T.; Kimmel, M.; Alphonse Ignatius, A. Probing conformational diversity of fc domains in aggregation-prone monoclonal antibodies. *Pharm. Res.* **2018**, *35*, 220. [CrossRef]
172. Nakamura, H.; Oda-Ueda, N.; Ueda, T.; Ohkuri, T. Introduction of a glycosylation site in the constant region decreases the aggregation of adalimumab fab. *Biochem. Biophys. Res. Commun.* **2018**, *503*, 752–756. [CrossRef]
173. Pepinsky, R.B.; Silvian, L.; Berkowitz, S.A.; Farrington, G.; Lugovskoy, A.; Walus, L.; Eldredge, J.; Capili, A.; Mi, S.; Graff, C.; et al. Improving the solubility of anti-lingo-1 monoclonal antibody li33 by isotype switching and targeted mutagenesis. *Protein Sci.* **2010**, *19*, 954–966. [CrossRef]
174. Wu, S.-J.; Luo, J.; O'Neil, K.T.; Kang, J.; Lacy, E.R.; Canziani, G.; Baker, A.; Huang, M.; Tang, Q.M.; Raju, T.S.; et al. Structure-based engineering of a monoclonal antibody for improved solubility. *Protein Eng. Des. Sel.* **2010**, *23*, 643–651. [CrossRef]

Review

Antibody Conjugates-Recent Advances and Future Innovations

Donmienne Leung [1,*], Jacqueline M. Wurst [2], Tao Liu [2], Ruben M. Martinez [2], Amita Datta-Mannan [3] and Yiqing Feng [4]

1 Biotechnology Discovery Research, Lilly Research Laboratories, Lilly Biotechnology Center, Eli Lilly and Company, San Diego, CA 92121, USA
2 Discovery Chemistry and Research Technology, Lilly Research Laboratories, Lilly Biotechnology Center, Eli Lilly and Company, San Diego, CA 92121, USA; jwurst@lilly.com (J.M.W.); liu_tao2@lilly.com (T.L.); Martinez_ruben_martin@lilly.com (R.M.M.)
3 Exploratory Medicine & Pharmacology, Lilly Research Laboratories, Lilly Corporate Center, Eli Lilly and Company, Indianapolis, IN 46225, USA; datta_amita@lilly.com
4 Biotechnology Discovery Research, Lilly Research Laboratories, Lilly Technology Center North, Eli Lilly and Company, Indianapolis, IN 46221, USA; feng_yiqing@lilly.com
* Correspondence: leung_donmienne@lilly.com

Received: 4 December 2019; Accepted: 21 December 2019; Published: 8 January 2020

Abstract: Monoclonal antibodies have evolved from research tools to powerful therapeutics in the past 30 years. Clinical success rates of antibodies have exceeded expectations, resulting in heavy investment in biologics discovery and development in addition to traditional small molecules across the industry. However, protein therapeutics cannot drug targets intracellularly and are limited to soluble and cell-surface antigens. Tremendous strides have been made in antibody discovery, protein engineering, formulation, and delivery devices. These advances continue to push the boundaries of biologics to enable antibody conjugates to take advantage of the target specificity and long half-life from an antibody, while delivering highly potent small molecule drugs. While the "magic bullet" concept produced the first wave of antibody conjugates, these entities were met with limited clinical success. This review summarizes the advances and challenges in the field to date with emphasis on antibody conjugation, linker-payload chemistry, novel payload classes, absorption, distribution, metabolism, and excretion (ADME), and product developability. We discuss lessons learned in the development of oncology antibody conjugates and look towards future innovations enabling other therapeutic indications.

Keywords: antibodies; site-specific conjugation; bioconjugates; ADC; antibody-drug conjugates; payloads; linkers; nucleic acids; ADME; developability; formulation

1. Introduction

Since the first monoclonal antibody drug approval (OKT3) in 1986, over 60 antibody therapeutics have become marketed drugs to date [1]. The number of protein therapeutics entering clinical development, including antibodies, antibody fragments, bispecifics, Fc-fusion proteins, and antibody-drug conjugates is expected to grow due to robust pipelines and high success rates for treating various diseases [2,3]. With the advances and extensive experience in antibody engineering over the past decades [4], antibody therapeutics have evolved from murine (e.g., OKT3) to chimeric (e.g., Rituxan®) to fully human (e.g., Humira®) as depicted in Figure 1. Monoclonal antibody-based therapeutics have been built to deliver specific effector functions or as bispecifics and conjugates to achieve the desired pharmacological effects [5,6]. Antibody discovery was enabled by murine hybridoma technology [7] followed by humanization [8] to deliver therapeutic antibodies with lower risk of immunogenicity [9]. Display technologies and transgenic animals have pushed the boundaries to produce antibodies with fully human sequences [10]. Antibody conjugates have similarly taken advantage of the progress made in monoclonal antibody development and improvements in conjugation chemistries [11–13] to expand the druggable target space for antibody-based therapies. These advances in antibody development are crucial to the success of antibody conjugates.

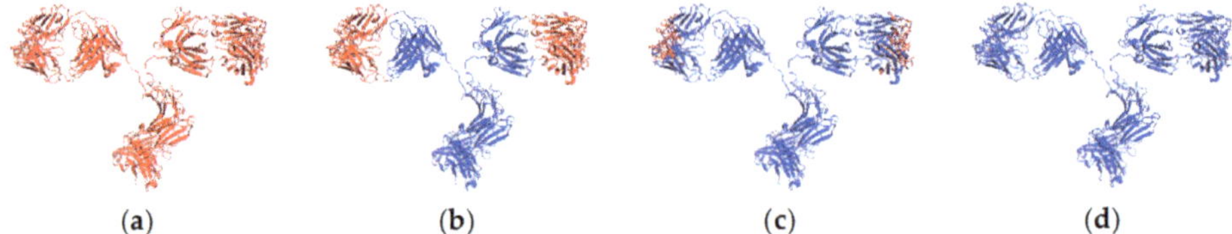

Figure 1. The evolution of (**a**) murine, (**b**) chimeric, (**c**) humanized, and (**d**) fully human monoclonal antibodies through protein engineering. Red and blue represents mouse and human antibody sequence respectively. The antigen binding complementarity determining regions (CDRs) are shown as sticks. The new generation of antibody-drug-conjugates (ADCs) utilized humanized (**c**) and fully human antibodies (**d**).

While chemotherapy and radiation have been the dominant treatments of cancer for decades, their lack of ability to distinguish between healthy and tumor cells has fueled the desire to create tumor specific delivery of cytotoxic payloads and radionuclides via antibody conjugates. Oncology antibody conjugates have successfully delivered potent chemotherapeutic and radioactive agents to kill tumor cells [12]. Currently, all of the FDA approved antibody-drug-conjugates (ADCs) are targeted cancer therapies (Table 1) [14], including the latest approval in June 2019 for Polivy® [15]. Herein, we review the progress made in oncology ADCs in terms of conjugate design and development, linker payload conjugation chemistries and highlight novel non-oncology conjugate innovations.

2. Critical Considerations for Antibody Conjugates

First generation oncology ADCs in the 1990s were based on murine or chimeric antibodies which were plagued with immunogenicity issues [16] and linker instability [17]. Immunogenicity of protein therapeutics has a critical impact on the pharmacokinetics and drug disposition and ultimately clinical success [18,19]. These molecules were designed to deliver a variety of protein toxins [20] and microtubule binding drugs [21] as the cytotoxic payloads. Limited antigen density on tumors, low potency of the payloads, and the low average drug-antibody-ratio (DAR ~3–4) prevented efficacious quantity of drug delivered, which was proposed to be one of the reasons for initial ADC failures, while higher DAR conjugates suffered from toxicity and low therapeutic index. Second generation ADCs from the last 10+ years approved by the FDA were armed humanized antibodies coupled to stabilized linkers and more potent payloads, such as auristatins, calicheamicins, and maytansinoids (Table 1).

In an ideal situation, ADC payloads should be inactive in circulation when conjugated to an antibody via a linker and remain stably conjugated until the conjugate reaches the target of interest. Upon internalization of the conjugate-target complex, active payload is released inside target cells after lysosomal degradation of the linker or the antibody itself. In addition to reducing target-independent uptake, conjugate stability remains crucial for specific delivery and distribution of payload to the target tissue from systemic circulation. Conjugation sites, chemistries and linker designs coupled with DAR load greatly affect plasma stability, biophysical properties, and consequently pharmacokinetics of the conjugate. Next generation ADCs will likely incorporate fully human antibodies with site-specific conjugation and novel linkers to reduce immunogenicity and optimize biodistribution and payload delivery. These topics will be discussed in this review.

Table 1. FDA approved antibody-drug-conjugates (ADCs).

ADC Product	Indications	Approval Date	Target Antigen	Antibody Conjugation	Average Drug Antibody Ratio (DAR)	Linker	Payload
Mylotarg® (Gemtuzumab ozogamicin)	Relapsed AML	2001, withdrawn 2010; (reapproved 2017)	CD33	Humanized IgG4—lysine	2–3	Hydrazone	Calicheamicin
Adcetris® (Brentuximab vedotin)	Relapsed HL and sALCL	2011	CD30	Chimeric IgG1—cysteine	~4	Dipeptide cleavable (Val-Cit)	Monomethyl auristatin E (MMAE)
Kadcyla® (Trastuzumab emtansine)	HER2 + metastatic breast cancer	2013	HER2	Humanized IgG1—lysine	3.5	Thioether Non-cleavable	Emtansine (DM1)
Besponsa® (Inotuzumab ozogamicin)	Relapsed or refractory CD22 + B-ALL	2017	CD22	Humanized IgG4—lysine	~4	Hydrazone	Calicheamicin
Polivy® (Polatuzumab vedotin)	Relapsed or refractory DLBCL	2019	CD79b	Humanized IgG1—cysteine	3.5	Dipeptide cleavable (Val-Cit)	Monomethyl auristatin E (MMAE)

2.1. Target and Antibody Selection

One of the key contributing factors to clinical failures has been the bio-distribution of an ADC, which is critically dependent on the relative target expression as well as target-independent uptake. Other aspects such as conjugate and linker stability and payload properties are described in other sections. Preferably for an oncology treatment with a biologic, antigen targets should have high expression levels on tumor cells and little to no expression on normal tissues. Internalization of the target-antibody complex is crucial for specific intracellular release of payloads. Antibodies are ideal delivery vehicles due to their high specificity to targets and long half-life, which is the result of pinocytosis and subsequent neonatal Fc receptor (FcRn)-mediated recycling [22]. Prolonged systemic circulation enables conjugate accumulation at the target sites.

The antibody Fc choice is an important consideration for both monoclonal antibody therapeutics and ADC therapeutics [23]. Fc-mediated effector functions such as antibody-dependent cellular cytotoxicity (ADCC) or complement-dependent cytotoxicity (CDC) are part of the mechanism of action for depleting antibodies [24]. However, with ADCs, the contribution of effector functions to efficacy and toxicity are not well understood. It is noted that two out of the five currently FDA approved ADCs (Mylotarg® and Besponsa®) employed IgG4 antibodies which lack effector functions. Although the effector functions have the potential to augment the anti-tumor activities of the ADCs, engaging Fcγ receptors is also a possible cause for off-target and dose-limiting toxicity (reviewed in [25].) Emerging literature suggests that the antibody internalization and delivery of the toxic drug to the target cells serves as the primary mechanism of action for ADCs that is far more efficient than ADCC and Antibody-Dependent Cellular Phagocytosis (ADCP). For example, trastuzumab-DM1, SYD985, and DS-8201a all target HER2 and have shown similar ADCC activity as trastuzumab but they have demonstrated dramatically more anti-tumor activity than trastuzumab [26–31]. The anti-Trop-2 ADC IMMU-132 represents a more striking case as this ADC lost 60%–70% of the ADCC activity compared with the unconjugated mAb upon the conjugation of SN38 [32]. Nevertheless, this ADC demonstrated significant antitumor effects in mice bearing human pancreatic or gastric cancer xenografts [32] and is showing promise in clinical trials [33,34]. On the other hand, it is well established that afucosylated IgG1 increased binding to FcγRIIIa on effector cell such as natural killer cells and led to enhanced ADCC activity [35]. An example is GSK2857916, an afucosylated IgG1 antibody as a non-cleavable MMAF conjugate targeting B-cell maturation antigen (BCMA), in the clinic for multiple myeloma and demonstrated potent anti-tumor activity while it harnessed multiple cytotoxic mechanisms [36,37].

Due to the large size of an antibody conjugate, the stromal barrier [38] and tumor tissue penetration is an obvious obstacle for oncology ADCs to overcome for the treatment of solid tumors. Nonetheless, the successful targeting and delivery of payload to liquid tumor in circulation pushed the concept and led to a new frontier in ADCs to deliver small molecules for non-oncology indications [39,40]. Novel non-cytotoxic payloads have been conjugated to antibodies in the hopes of extending the pharmacokinetic properties and increasing therapeutic index of the drugs. Genentech has pioneered antibody-antibiotic conjugates [41,42] to target intracellular *Staphylococcus aureus* within host cells. Others have leveraged the internalization mechanism of antibodies to deliver immunosuppressive, cardiovascular or metabolic disorder small molecule drugs to specific cells using cell surface targets such as E-selectin [43], CD11a [44,45], CD25 [46], a3(IV)NC1 [47], CXCR4 [40,48], CD45 [49], CD70 [50], CD74 [51], and CD163 [52,53]. Examples of linker payloads as well as formulation and delivery challenges for non-oncology indications are discussed below. Additionally, genes of interest have been targeted in specific cell types to produce durable response using antibody-oligonucleotide conjugates [54,55]. Delivery of oligonucleotides have traditionally been challenging and various modifications have been employed to facilitate better cell penetration. This is explored in a later section.

2.2. Conjugation Methods

Antibody conjugation methods (Figure 2) have been extensively reviewed [11,56–58]. To date, all the FDA approved ADCs have relied on coupling reactions using either the nucleophilic primary amino group of surface-exposed lysines or the thiol group of reduced structural disulfides. The resulting product is a controlled heterogeneous mixture of antibodies with average drug load. High DAR species leads to aggregate formation, lower tolerated dose, and faster systemic clearance while low DAR species suffer from low efficacy [59]. Although DAR profile can be controlled by conjugation process development and specific DAR can be purified, site-specific methods to produce more homogeneous drug products would improve yield and biophysical properties, which will be critical for the next generation of ADCs. Towards these ends, extensive experience in protein engineering has allowed strategic placements of residues at specific locations enabling chemo-selective conjugation reactions. Researchers at Genentech first demonstrated that conjugation stability is location dependent and specific engineered cysteine sites were able to improve therapeutic index [60–62]. Cysteine insertions at specific sites can also efficiently produce stable conjugations [63]. Others have shown similarly that location of the conjugation sites can impact the stability and pharmacokinetics of the ADCs using alternative residues and chemistries [64,65].

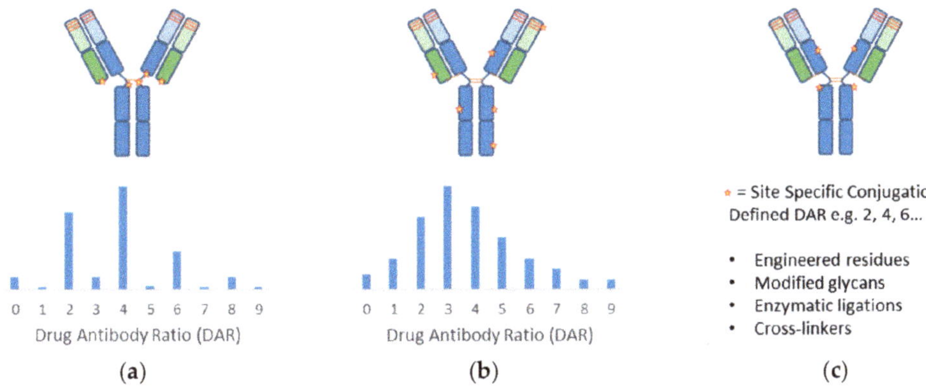

Figure 2. Antibody conjugation methods include (**a**) cysteine-reactive, and (**b**) lysine-reactive chemistries which generate heterogeneous mixtures of drug-antibody-ratio (DAR), while (**c**) site specific conjugation methods deliver more homogeneous product with defined DAR using engineered residues, modified glycans, enzymatic ligations, and chemical cross-linkers. Schematic representation of antibody heavy chains and light chains are colored blue and green respectively. complementarity determining regions (CDRs) and conjugation sites are depicted as red bars and stars respectively. Approximate DAR distribution for stochastic cysteine and lysine conjugations are presented as bar charts.

Enzymatic methods have also been explored (reviewed in [66]) where recognition sequences have been engineered into the antibody to facilitate site-specific conjugation. Most well-exemplified in this category are enzymes such as transglutaminase [65,67–69], sortase [70–72] and formylglycine-generating enzyme (FGE) [73,74]. Transglutaminases (TG) catalyze a stable isopeptide bond between an amine of a lysine and the γ-carbonyl amide of a glutamine. Deglycosylation of N-linked glycan on a native antibody exposes glutamine at position 295 for site-specific conjugation with TG either through direct coupling with an amine-functionalized linker payload or via a two-step coupling by installing bio-orthogonal azide or thiol for strain-promoted azide-alkyne cycloaddition and maleimide chemistry respectively [67]. Alternatively, glutamine residues can be engineered and short glutamine (LLQG) tags were introduced into different regions to yield highly stable site-specific conjugates with good pharmacokinetic profiles [65,68,69]. Sortase catalyzes a transpeptidation reaction between a N-terminal glycine of GGG peptide or linker payload with the threonine-glycine bond in a LPXTG motif to produce a peptide fusion or site-specific ADC with high in vitro and in vivo potency [70–72]. Lastly, SMARTag® [75] is an example where formylglycine-generating enzyme (FGE) converts an engineered cysteine residue in a specific peptide sequence to produce an aldehyde tag in cell culture [73,74] to enable conjugation with linkers via oxime formation or a Pictet–Spengler reaction [76,77].

Other conjugation chemistries involved the engineering of unnatural amino acids [78–82] to install reactive groups in the antibody for bio-orthogonal chemistry [77]. Companies such as Ambrx [81] and Sutro Biopharma [83] have utilized these elegant approaches to generate site-specific ADCs. An orthogonal amber suppressor tRNA/aminoacyl-tRNA synthetase pair is used to incorporate the unnatural amino acids such as para-acetylphenylalanine (pAF), para-azidophenylalanine (pAZ), and para-azidomethylphenylalanine (pAMF) into recombinantly expressed antibodies in cell-based or cell-free systems. Reactive ketone in pAF forms a stable oxime linkage with alkoxyamine containing linkers, while the azido group in pAZ and pAMF undergoes click chemistry with alkynes to produce homogenous ADCs.

Alternatively, native antibodies can be conjugated site-specifically [84] after enzymatic modification of natural amino acids such as tyrosines [85] and glutamines [67], and carbohydrates can be oxidized chemically or modified with enzymes to produce reactive groups for conjugation [86–88]. For instance, sodium periodate oxidation of fucose at the native N-linked glycan of an antibody installed an aldehyde for conjugation with hydrazides [86]. Glyco-remodeling methods include enzymatic transfer of galactose and sialic acid with a mixture of transferases yielded glycans which can be oxidized with periodate to form oxime conjugates with aminooxy linker payloads [87], while sialytransferase can also incorporate an azide modified sialic acid derivative into the antibody for click chemistry [89]. Similarly, SynAffix BV utilized a 2-steps GlycoConnect™ process to trim a mixture of glycoforms with endoglycosidase, followed by enzymatic transfer of azido sialic acid for copper-free click chemistry [88]. Lastly, site-specific conjugation approach to retain the structural stability of a native antibody is to cross-link the reduced interchain disulfides with re-bridging chemical reagents [90–94].

Taken together, the various novel site-specific conjugation methods often require additional investments in manufacturing processes to produce the clinically viable products at large scale. Many development advances have been made in site-specific conjugations enabled a new generation of ADCs to enter the clinic in recent years, but this topic is outside the scope of this review.

3. Current Small Molecule Payloads and Beyond

ADC payloads have been mostly anti-mitotic small molecules for oncology indications [13,39]. The clear advantage of conjugates in this space is targeted delivery at efficacious doses that are below what could be given systemically due to the toxicity of the payload. Current approved ADCs such as Adcetris®, Kadcyla®, and Polivy® carry cytotoxic payloads that rely on the anti-mitotic mechanism of action (MOA) such as monomethyl auristatin E and emtansine (Table 1; Figure 3a,b, respectively). This class of payloads still predominates in clinical stage research, including novel derivates of these small molecules, despite efforts to use payloads with alternative MOAs [13,39,95].

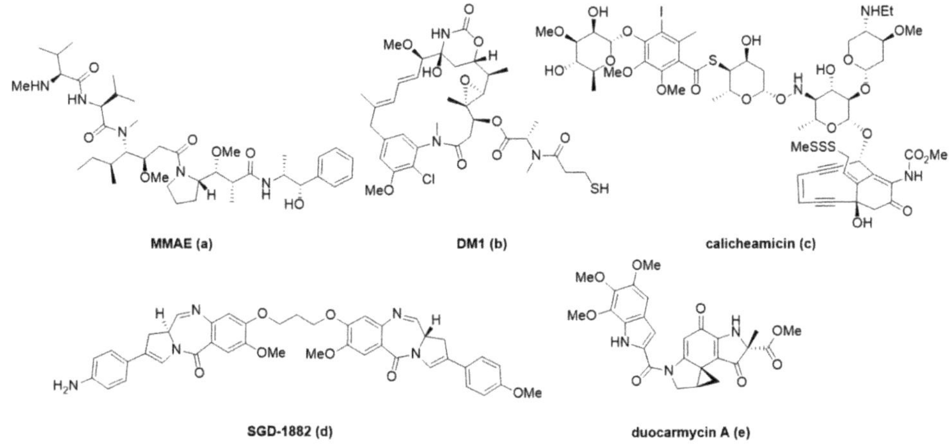

MMAE (a)

DM1 (b)

calicheamicin (c)

SGD-1882 (d)

duocarmycin A (e)

Figure 3. Examples of ADC payloads used clinically include, monomethyl auristatin E (MMAE, **a**, emtansine (DM1, **b**), calicheamicin (**c**), pyrrolobenzodiazepine dimer (PBD, SGD-1882, **d**), and duocarmycin A (**e**).

DNA damaging agents are another class of well-studied payloads [96–98]. Enediynes, such as calicheamicin **c**, are the warhead in Mylotarg® and Besponsa® (Table 1, Figure 3). They act through DNA-binding and induction of DNA-double strand breaking to produce a cytotoxic response in target cells. Other important DNA damaging agents currently used as payloads in clinical trials are duocarmycins **e** and pyrrolobenzodiazepines (PBD, **d**); they have unique, well-understood minor groove binding mechanisms that disrupt normal DNA function leading to cell death. Novel payloads have been explored recently in this space such as bis-intercalator depsipeptides with nanomolar affinity to DNA [99]. Pfizer demonstrated that this ultra-potent payload **a** (Figure 4) can overcome the previous limits of efficacy in animal models, therefore, expanding their relevance to indications beyond liquid tumors.

Further, highly-potent payloads described in the conjugate space (Figure 4) are pyrrole-based KSP (kinesin spindle protein) inhibitors **b** [100], which are traditional in the sense of their antimitotic mechanism-of-action, but newly incorporated pyrrole functionality provided an increase in efficacy against a wide-range of cancers previously untouched by this class. Continuing the push beyond the limits of first-generation payloads, Daiichi Sankyo® has incorporated a topoisomerase I inhibitor, exatecan derivative (DXd, **c**) into an ADC (DS-8201a). With superior pharmacodynamic and safety properties, in large part from the payload, DS-8201a has shown promising response in trastuzumab emtansine-insensitive cancers [31].

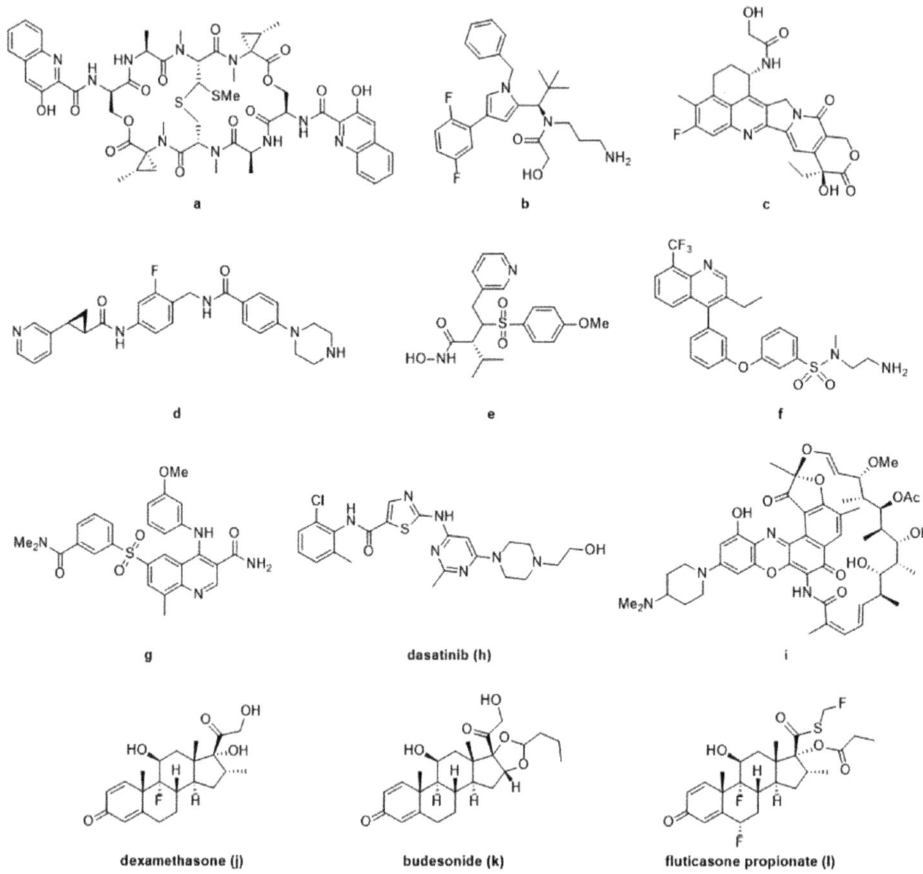

Figure 4. Expanding payload space in oncology with DNA disrupting bis-intercalator depsipeptide (SW-163D, **a**), pyrrole-based kinesin spindle protein (KSP) inhibitor (**b**), topoisomerase I inhibitor (DXd, **c**), nicotinamide phosphoribosyltransferase (NAMPT) inhibitor (**d**), and MMP9 inhibitor (CGS27023A, **e**). Examples of non-oncology payloads include LXR agonist (**f**), PDE4 inhibitor (GSK256066, **g**), kinase inhibitor dasatinib (**h**), antimicrobial rifamycin analog (**i**), GR agonists dexamethasone (**j**), budesonide (**k**), and fluticasone propionate (**l**).

As with other oncology examples, nicotinamide phosphoribosyltransferase (NAMPT) inhibitors **d** have not succeeded in the clinic due to low therapeutic index with limiting toxicities, but exploitation of ADC targeted delivery of NAMPT inhibitors provides an outlet for these potent payloads [101] in preclinical studies. Similarly, an anti-MMP9 antibody was conjugated to a non-selective MMP inhibitor (CGS27023A, **e**), showing remarkable selectivity for MMP9 alone, due to the antibody targeting in vitro [102].

Beyond oncology, ADCs are just beginning to show promise as a novel modality, offering a potential solution to the pitfalls of traditional drug discovery. Using an anti-CD11a antibody conjugated to an LXR (Liver X Receptor) agonist **f**, researchers were able to target macrophages to reverse cholesterol transport and reduce inflammation without negatively affecting hepatocytes, which have previously shown on-target toxicity with LXR agonists [44]. Antibody targeting CD11a was also used to selectively deliver a known PDE4 (Phosphodiesterase 4) inhibitor (GSK256066, **g**) and reduce inflammatory cytokine production, showing promise for the treatment of chronic inflammatory conditions with a more optimal therapeutic index [45]. In a similar fashion, dasatinib **h**, a known Src-family kinase inhibitor against leukemia was re-purposed as an immunosuppressive ADC using an anti-CXCR4 antibody to specifically target T-cells without undesirable side-effects [48].

Antimicrobial research is another area demanding novel approaches; it is well established that our current arsenal against microbes is failing and the discovery of new molecules has been limited [103]. Genentech pioneered the antibody-antibiotic conjugate (AAC) to deliver a rifamycin analog **i** intracellularly via an anti-*S. aureus* antibody which demonstrated marked clearance of latent bacteria reservoirs thought to be the cause of recurring infection [41].

Other known classes of molecules, namely, glucocorticoids are used as a standard of care for many immunological indications. They come with less-than-desirable side-effects at efficacious doses [104,105] with chronic use. One of the first examples of targeted glucocorticoid delivery via antibody used an anti-E-selectin conjugate to deliver dexamethasone to TNFα stimulated endothelial cells [43]. These early proof-of-concept experiments tracked conjugate internalization, intracellular release of steroid, and reduction of the pro-inflammatory IL-8 expression. Expanding on these preliminary experiments, an anti-CD163 conjugate was used to deliver dexamethasone to macrophages showing a synergistic anti-inflammatory effect of the conjugate versus its components alone, and a significant reduction in the systemic steroidal side-effects of orally dosed dexamethasone at the same efficacy [52]. Known glucocorticoid receptor (GR) agonists (i.e., dexamethasone **j**, budesonide **k**, and fluticasone propionate **l**) have also been attached to anti-CD74, anti-CD70, and anti-CD25 antibodies showing an immune cell targeted anti-inflammatory response, as well as highlighting the complexities of developing ADCs in this therapeutic area [46,50,51,106]. This novel modality promises treatment to larger populations of patients with autoimmune disorders, in a disease specific fashion that could potentially replace traditional steroid treatments as the standard of care.

4. Nucleic Acid Conjugates

For traditional oncolytic ADCs, the challenge of delivery manifests in the form of systemic toxicity of the potent cytotoxic payloads. However, similar molecular calculus may be applied to extensively cleared molecular entities which may otherwise have trouble reaching the tissues of interest. A rapidly-advancing molecular space which is typically impeded by delivery issues is that of synthetic therapeutic oligonucleotides including antisense oligonucleotides (ASOs) and short interfering RNA (siRNA) [107–109].

While traditional ADC payloads are small molecules which act on cellular machinery to elicit their desired phenotype via action on a molecular target, the current generation of synthetic oligonucleotide therapeutics instead act on information molecules upstream of their targets such as endogenous mRNA, achieving specificity through base-pair complementarity. Several different mechanisms of action have been clinically validated. Examples include intronic splice modulation (i.e., nusinersen, an ASO for treating spinal muscular atrophy) and formation of a catalytic mRNA silencing complex (i.e., patisiran, a lipid-nanoparticle formulated siRNA for treating ATTR amyloidosis). The details of these mechanisms, along with others, have been recently reviewed [110] and are beyond the scope of this discussion. The chemical structures and properties of oligonucleotides that make this therapeutic space challenging, however, are central to this discussion; their poor tissue penetration and circulation half-life are often attributed to the polyanionic backbone characteristics [111,112].

A number of advances in both oligonucleotide chemistry as well as conjugate chemistry have been enabling oligonucleotide clinical candidates as a class and will likely impact oligonucleotide bioconjugates [113]. For example, it has been demonstrated that appending of N-acetylgalactosamine (GalNAc, Figure 5b) residues to the end of siRNA strands allows for efficacious loading of hepatocytes in vivo with sustained target knockdown [114]. Along with GalNAc, number of other common modifications in oligonucleotide chemistry have been utilized, such as 2′-modification (i.e., 2′-F and 2′-OMe nucleosides, Figure 5a) and sulfurized phosphate analogs (i.e., phosphorothiolates, Figure 5a). Similar themes of heavy chemical modification have been utilized in ASOs. For example, one report principally noted that GalNAc conjugation ameliorated some nephrotoxicity signals in PCSK9-targeting ASOs, with data suggesting such conjugation could be a path towards addressing oligo-induced nephrotoxicity concerns more generally [115]. Beyond GalNAc, other oligonucleotide conjugation strategies have been found effective pre-clinically in targeting specific cellular populations or promoting systemic availability, as with conjugation of a GLP1R agonist [116] or conjugation to various lipids such as cholesterol or docosahexaenoic acid (DHA, Figure 5b) [117], respectively.

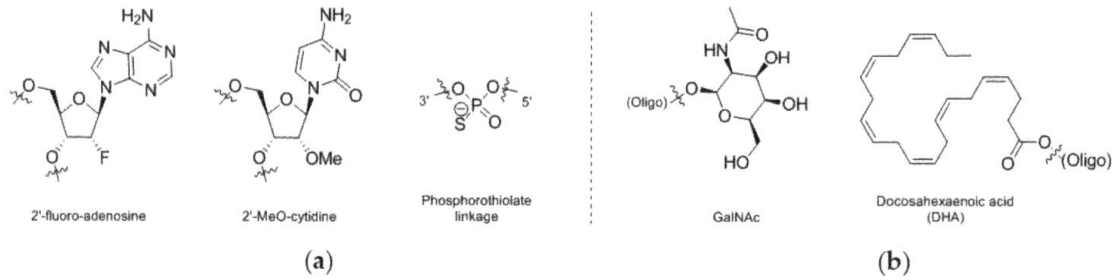

Figure 5. (**a**) stabilizing chemical modifications for siRNA, (**b**) oligo delivery conjugate moieties.

Although there are limited examples of discreet therapeutically-oriented antibody-oligonucleotide conjugates in the literature, the current studies have made significant headway and put to use many of the strategies discussed herein. In an early example reported, hu3S193, an internalizing humanized Lewis-Y mAb was conjugated to a largely unmodified STAT3 siRNA [118]. This conjugation by utilizing non-specific amino residue labeling of hu3S193 with activated hydrazonal nicotinamide (HyNic) reagent which was covalently coupled with STAT3 siRNA using an aldehyde linker. Cellular specificity was confirmed through flow cytometry and internalization by confocal microscopy, the construct could only effect STAT3 knockdown when high doses (100 μM) of chloroquine as an endosomal disrupting agent was added.

The disconnect between specific internalization and siRNA target knockdown was further demonstrated in a systematic study which applied a number of important antibody-drug conjugate parameters [119], which included variations of linker chemistry, cell surface receptor identity, receptor internalization types and count, antibody linkage positions, and antibody formats. This study utilized several sophisticated methods, including the application of site-specific engineered cysteine conjugation (Genentech's THIOMAB™ platform) to deliver reproducible and largely homogeneous conjugates, as well as highly modified chemically-stabilized housekeeping gene (PPIB) siRNA constructs to enable in vivo studies. While seven different internalizing antigens were profiled, only three showed any knockdown of the target gene and of those only TENB2 on cell lines with high surface receptor density was able to achieve knockdown levels of greater than 50% in vitro. It was inconclusive what properties of the different active conjugates enabled effective knockdown, but importantly, conjugates were able to demonstrate 33% transcript knockdown with cellular specificity to tumor cells near the vasculature in a mouse xenograft model.

At this time, there have yet to be any clinical studies on antibody-oligonucleotide conjugates, but several of the requisite preclinical proof-of-concept studies have been reported. One notable example has demonstrated application of a myostatin-silencing siRNA-antibody conjugate in a mouse model of muscular regeneration [120]. In this study, the well-profiled CD71 receptor (transferrin receptor) was chosen for muscular target engagement utilizing anti-CD71 Fab′-siRNA conjugates. In profiling methods of administration, principal findings were that equivalent target engagement was achievable through different perfused systemic administration routes. PCR-based detection methods were used to verify conjugate was detectable 24 h post-dose, but notably durable silencing up to a month out was observed. Additionally, intramuscular injection enabled superior levels of target engagement in a model of peripheral artery disease, with muscular regeneration due to myostatin knockdown observed with microgram-scale injections. While this Fab′-based study utilized structural cystines for conjugation, other therapeutic antibody-oligonucleotide conjugation systems not yet described in this section have been reported, including two oncology examples which applied two-step conjugations for azide-labeling the antibody which were then treated with cyclooctyne-appended ASOs to generate the conjugates [121,122]. Other antibody-oligonucleotide conjugation methods beyond the scope of this review (i.e., for immuno PCR applications, pre-targeted radiotherapy, or those which utilize non-covalent heterogeneous complexes) have been recently reviewed [109].

While these results are promising, further analytical work must demonstrate pharmacokinetic profiling of these conjugates to enable clinical study. In this vein, a report describing a triplex forming oligonucleotide ELISA assay utilized locked nucleic acid-containing probes [123] for conjugate quantification. This assay is distinguished from PCR-based methods as a direct, quantitative readout of intact oligonucleotide-antibody conjugate. The authors demonstrated that this assay is capable of accurately detecting conjugate doped into cellular matrices such as serum or tissue homogenate down to a limit of detection of 120 pg/mL. These early developments in antibody-oligonucleotide conjugates can help propel the next wave of novel conjugates to leverage the cellular specificity of antibodies and target-gene specificity of oligonucleotide therapies.

5. Linkers

While a simple concept at first glance, a linker is far more complex than a mundane spanning element between the small molecule payload and the antibody which make up the ADC. It ensures the fundamental principles of targeted drug delivery of ADCs-minimizing premature drug release in plasma and promoting selective release of payload to the target cell. Additionally, it can modulate the physiochemical property of the overall conjugate. This requires the linker design to be stable in circulation and upon antibody-mediated internalization, the payload is efficiently released.

To meet the desired therapeutic effect, cytotoxic payloads can be designed to be released either intracellularly or extracellularly [10,13,124–126]. For instance, intracellular hydrolytic enzymes can recognize specific linker motifs (Figure 6a) and catalyze the cleavage reactions to release payloads inside endosomes or lysosomes. Cathepsin family enzymes and other proteolytic enzymes are responsible for the digestion of peptide linkers in lysosomes [69]. Selection of the linker peptide sequence affects the plasma stability of ADCs and the efficiency of proteolytic cleavage of linkers in target cells [127,128]. Phosphatases [106] and glycosidases [129,130] are other examples of hydrolytic enzymes that are present in lysosomes at high concentrations and break down respective linkers of internalized ADCs. Alternatively, through exploitation of the relatively acidic and reductive tumor microenvironment, payloads may also be released extracellularly through non-enzymatic cleavable linkers (Figure 6b) at a tumor site. For example, hydrazine and acyl hydrazine (approved products in Table 1) [25,131], ketal and acetal [132], as well as carbonate [133] are acid-labile linkages that are designed to degrade at low pH in cell chambers (pH 5.5 in endosome, pH 4.5 in lysosome) and remain intact in circulation (pH 7.4). Disulfide linkages are subjected to glutathione attack in cytosol, thus offering a chance of selective releasing toxic payloads in tumor cell. How the conjugation site and steric hindrance around disulfide linkage modulates its stability in plasma is a subject that has attracted extensive studies [134,135].

Both enzymatic cleavable linkers and non-enzymatic cleavable linkers have been conjugated onto antibodies through naturally occurring cysteine and lysine residues (Figure 6c). Nucleophilic thiol groups can react with maleimide and alkyl-halide to produce stable conjugates. In cases of maleimide-based ADCs where stability is a concern, semi-hydrolysis of maleimide has been reported as an effective strategy to minimize the retro-Michael reaction of thiomaleimide and preventing premature payload loss [136]. Amino groups on surface exposed lysines can form stable amide via alkylation of an activated ester on the linker. Side chains of natural amino acids can be modified or engineered to produce ketones to form imines or hydrazones. Notably, recent site-specific ADCs utilize novel linker conjugation chemistries with unnatural amino acids that are incorporated on engineered antibodies through imine/hydrazine formation, copper-catalyzed azide alkyne cycloaddition (CuAAC), and strain promoted azide alkyne cycloaddition (SPAAC) [44,78,80].

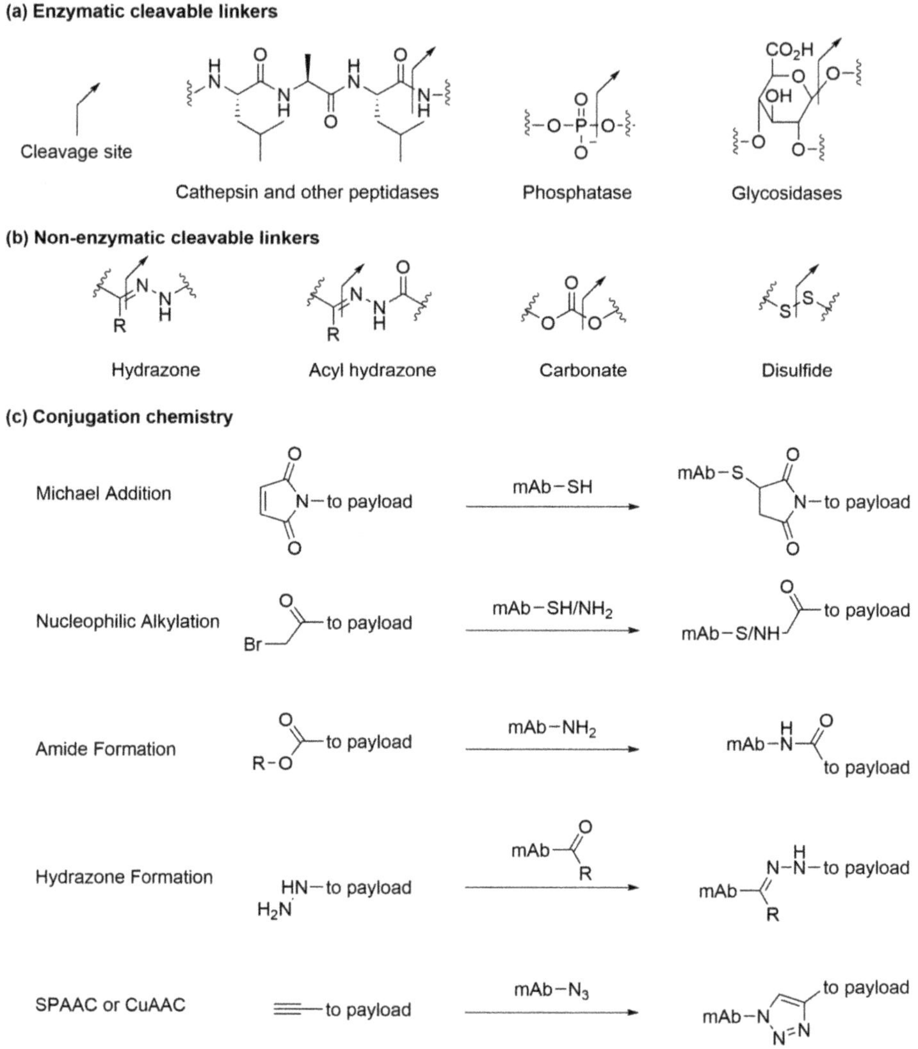

Figure 6. (**a**) Enzymatic cleavable linkers; (**b**) non-enzymatic cleavable linkers; (**c**) conjugation chemistry.

In sharp contrast to approved cytotoxic ADCs for oncological indications, the goal of ADCs for non-oncological applications is the selective modulation of target cells without on-target adverse bystander effects. Besides selection of non-toxin-based payloads, this new direction also demands new concepts in linker-payload design. Linker-payload design is critical in modulating bystander effect of payloads. A diphosphatase-cleavable linker has been reported for the selective delivery of immune suppressing payloads to immune cells following ADC internalization and diphosphate cleavage (Figure 7) [51]. In this study, the fluticasone propionate derived payload has a high intrinsic binding affinity to target, but also bears a charged phosphate moiety. Due to antibody-driven delivery and limited free payload permeability, the target exposure of payload to cell is increased, and a superior in vitro potency is observed. Further in vivo testing may require additional linker development.

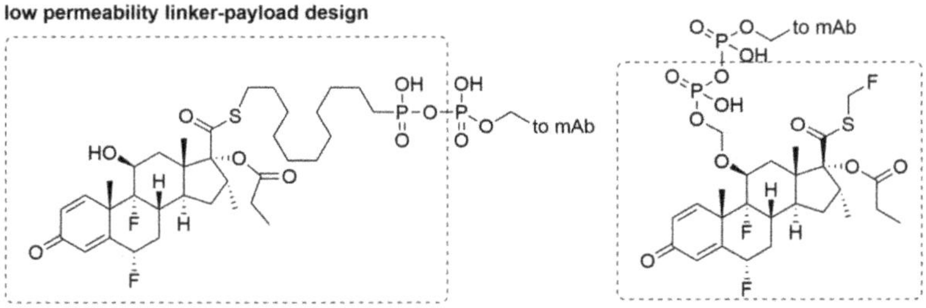

low permeability linker-payload design

Figure 7. Low permeability linker-payload design.

Linker-payload design is also critical to the successful implementation of high-DAR ADCs, via reduced aggregation and improved overall pharmacokinetics profiles. Even though increasing DAR instinctively increases in vitro potency of ADCs, it may not translate to an improvement of in vivo potency since the plasma clearance of ADCs rises along with DAR [129]. The aggregation and fast clearance problem may be largely mitigated without extensive linker optimization for water-soluble payloads such as the topoisomerase I inhibitor DXd in DS8201a, which is an ADC with DAR 8 that has shown remarkable in vivo stability both pre-clinically and clinically [31]. For highly hydrophobic payloads, the paradoxical effect of higher DAR resulting in lower exposure can be corrected by novel linker design. Several linker modifications to enhance hydrophilicity have been reported that allowed ADC to be produced with DAR as high as 8 (reviewed in [124]). These modifications include addition of a polyethylene glycol (PEG) moiety [127,128], a glucuronic acid unit [129], or a combination of branched PEG moiety and glucuronic acid unit [130]. For instance, hydrophilic linker construct in Figure 8 [129] minimizes the detrimental hydrophobicity associated with increasing DAR, imparting an optimal pharmacokinetic profile to the high DAR ADC, thereby reduces non-specific clearance and improves in vivo potency.

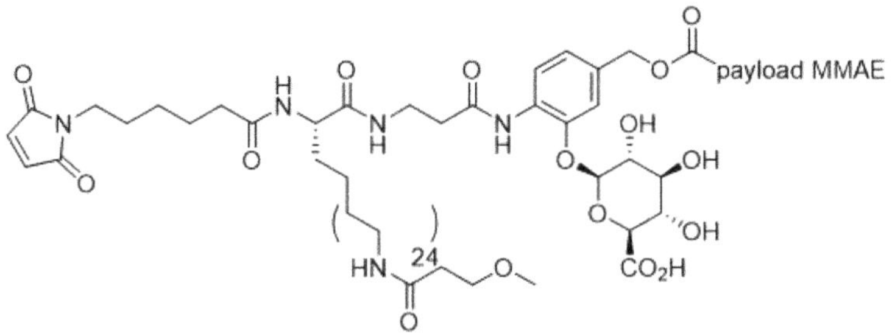

Figure 8. Hydrophilic linker-payload design.

6. Absorption, Distribution, Metabolism, and Excretion (ADME) of ADCs

As discussed in the sections above, considerable advancements in next generation ADCs are anticipated to further explore both the chemical and biological design elements for ADCs. This includes, but is not limited to, antibody engineering to facilitate direct site-specific conjugation, modification of conjugation chemistry and introduction of novel linker compositions to confer enhanced stability, as well as, the exploration of additional existing and novel chemical entities/modalities as conjugates to increase the pharmacological applications of ADCs. Many of these advancements to design better molecules are intricately interdependent with optimizing or improving the ADME drug-ability properties of ADCs, such as linker-payload stability, distribution, and pharmacokinetics (PK). This is because defining the exposure-response relationship for both safety and efficacy has been intimately tied with ADC peripheral PK, target tissue or site of action concentration and the disposition of the payload at the intended site. The criticality of understanding the exposure-response relationship and therapeutic index (TI) for ADCs is exemplified by Mylotarg® (gemtuzumab ozogamicin) which is composed of a CD33 mAb linked to the cytotoxic drug calicheamicin via an acid-liable hydrazine linker. While initially approved for the treatment of acute myeloid leukemia (AML) in 2000, it was pulled from the market at the request of regulatory agencies in 2010 due to safety concerns and the failure to reproduce the clinical benefit connected to linker stability in AML patients. Following additional interrogation of exposure-response relationships, examining alternative lower dosing and scheduling, Mylotarg found a path back to the market and received a new FDA approval for newly diagnosed CD-33 positive acute AML patients in 2017.

Dissecting the ADME properties of ADCs is a complex endeavor given the unique properties of each component within the molecules. ADC ADME involves delineating the intertwined properties of linker-payload stability, pharmacokinetics, clearance, metabolism, and disposition of mAb, conjugate (i.e., small molecule chemical entity for traditional ADCs), as well as, the ADC entity itself. The ADME properties of ADCs are influenced by the mAb, target antigen, linker, site of conjugation, DAR number, and the conjugated species (i.e., payload). Table 2 summarizes the various types of in vitro and in vivo studies to characterize ADC ADME.

Table 2. ADME Characterization approaches for ADCs and their constituents.

Species	ADME Information
Antibody	• Determine PK-dose relationship in vivo • Characterize target affinity/specificity in vitro, target expression/turnover in vivo, unintended or off-target binding in vitro and in vivo
ADC	• Linker Component (1) Characterize linker stability and kinetics of catabolism in vitro and in vivo across species (2) Evaluate nature of the released species (active payload and its catabolites) • Conjugation Site (1) Evaluate influence of conjugation site on linker stability in vitro and in vivo (2) Determine the effect of conjugation site on PK • DAR (1) Determine in vivo PK and disposition with heterogenous and homogenous DAR species
Payload	• Metabolite identification, characterize DDI potential (CYP inhibition, induction and reaction phenotypes) • P-gp substrate or inhibitor • Characterize non-P-gp transporters • Plasma protein binding

The antibody component of ADCs is the primary driver of the slow clearance, long systemic half-life and restricted tissue distribution of these modalities compared to their payload counterparts. Similar to mAbs, the properties related to target (expression pattern, density and turnover), as well as, antibody structure (including physiochemical properties, FcRn binding, Fcγ receptor interactions and isotype) that affect antibody PK and disposition also impact ADCs. An anti-drug antibody (ADA) or neutralizing antibody (Nab) response against the therapeutic antibody component can affect the PK profile and shorten the half-life of the ADCs in the body [9,19]. Nevertheless, idiotype networks have an important biological role in avoiding the expansion of autoreactive B or T-cells [137,138]. Uniquely, ADC PK is also impacted by linker composition, chemical nature of the payload and DAR. These are both speculated to affect the physiochemical properties of the ADC which are linked to the PK and clearance of the molecules. For example, ADCs with high DAR values have been shown to aggregate and have higher clearance rates than their unconjugated mAb counterparts or lower DAR species [59,129,139]. Similarly, decreasing the hydrophobicity and improving the hydrophilicity of the linker component within an anti-CD70 and anti-HER2 mAbs improved the exposure and changed the disposition of the ADCs [129,140].

In addition to the antibody, the linker and payload components of ADCs are also subject to their own clearance mechanisms. The stability of the linker to premature release of the payload in the systemic circulation has been demonstrated to be a critical ADC ADME component for determining the exposure-response and exposure-toxicity relationships [141]. From a stability perspective, well-behaved ADCs should only release the payload in the intended target tissue to minimize payload toxicity to unintended tissue and maximize efficacy in target tissues/organs, especially with payloads with cytotoxic properties. The ADC linker stability is noted to be a challenge due to the long circulating half-life (days to weeks) imparted by the mAb component resulting in the continuous assault of the linker to endogenous proteases. Mechanistically, linker stability can be evaluated both in vitro using plasma/serum incubations and in vivo following administration to multiple species by following the formation of the released payload and DAR changes over time. As covered above, linker composition continues to be an intense area of focus in the development of ADCs. In terms of the payload, initial reports with limited chemical entities suggested that type of payloads did not impact the PK of ADCs; however, more recent studies of conjugation sites and of site-specific conjugations have demonstrated the connectivity of the site of conjugation with various payloads to impact ADC PK and disposition [142,143]. Engineering ADCs for site-specific conjugation to control the DAR and PK has shown some evidence of improving the TI in non-clinical oncology studies [144]. Approaches such as engineered cysteines, unnatural amino acids, and the inclusion of tags (i.e., selenocysteine, aldehyde, or glutamine) continue to be intense areas of research for the application of site-specific payload conjugation to optimize ADC ADME.

Like mAbs, ADCs are likely trafficked via the vascular and lymphatic systems. The biodistribution of ADCs follows that of the antibody component. ADC are removed from the systemic circulation by target-mediated drug disposition (TMDD); thus, highly vascularized organs or tissues that express the target antigen are involved in the clearance of ADCs from the periphery. The TMDD is believed to be followed by intracellular trafficking of the target: ADC complex to lysosomes where degradation of the ADC occurs, and the payload is released from the mAb to elicit its activity. In addition, nonspecific uptake of ADCs by pinocytosis also facilitates their systemic depletion. This form of uptake could lead to degradation and/or recycling of the ADC by the neonatal Fc receptor (FcRn). Indeed, in terms of tissue distribution the preponderance of ADCs are observed in four organs including the liver, kidneys, lungs, and skin [145,146]; however, the amount in each tissue differs between ADCs based on their target binding and physiochemical properties [147]. Importantly, irrespective of the mode of ADC degradation, the payload or chemical moiety can be released into the blood. The unconjugated payload is expected to follow the biodistribution pattern of a typical small molecule drug which is widely distributed throughout the tissues.

The elimination of ADCs involves two processes. First, intracellular catabolism through proteolysis in the tissues (i.e., TMDD- or pinocytosis-mediated). Second, complete deconjugation of the payload which can result in both mAb or mAb with a partial linker along with free drug [148]. While the mAb based species are expected to follow catabolism through the same mechanisms as the ADC, there is increased attention in the elimination of the unconjugated payload under conditions of impairment of renal and hepatic processes. For example, a study of brentuximab vedotin showed that the major route of MMAE excretion was through the feces (~72%) and the remaining MMAE was recovered in urine in humans [149]. Given these data, the relationship of hepatic and renal insufficiency to ADC exposure is an important aspect of clinical development.

Another area of intense research is dissecting the noted phenomenon of resistance against ADCs. A few mechanisms of resistance have been noted including to the antibody portion of the ADCs by mutation and/or down-regulation of the target antigen, as well as, to the payload via drug efflux transporters that remove the payload from cells [150]. Changes in the intracellular processing of ADCs through alterations of the linker cleavage caused by lysosomal or endosomal abnormalities can also significantly affect the PK profiles. These changes impair the release of payloads in the cytosol and consequently affect the therapeutic indexes of the ADCs [150].

7. Conjugate Developability, Formulations, and Characteristics

An antibody conjugate combines an inherently complex antibody with a small synthetic molecule drug to create an even more complex large molecule. Despite the relatively small addition in molecular weight, the small molecule drug has a profound impact on the characteristics and properties of the conjugate.

Over the last decade, significant advancements in analytical methods have been made to characterize ADCs and have been extensively reviewed [151–154]. These methods have focused on the major ADC attributes such as DAR, drug load distribution, residual linker-payload and related impurity levels, in addition to typical attributes for antibodies such as aggregation level, charge variants, and host cell protein level. The DAR number has a strong influence on the properties of ADCs. Currently, many of the ADCs in the clinic have DAR numbers in the range of 2–4, although ADCs with higher DAR numbers have also been reported [31,155–158]. For stochastically conjugated ADCs, the small molecule drug is covalently linked to either the lysine or the interchain cysteine residues of the antibody, resulting in a heterogeneous mixture with various DAR species which are more difficult to characterize and control. For example, as many as 40 lysines were found to be partially modified in a lysine conjugated ADC molecule using LC-MS and peptide mapping methods [159]. The aforementioned site-specific conjugation approaches have drastically improved the DAR homogeneity albeit process development is required to further control the remaining heterogeneity during the production process [160]. At present, nearly all the antibody conjugate characterization literature has focused on antibody-small organic molecule conjugates. Based on the molecular nature of each payload, chromatographic and electrophoretic methods as well as spectroscopic methods have been commonly employed in DAR determination along with mass spectrometer method which provides more detail. The analytical methods will continue to evolve as the payload expands to nucleic acids which possess very different properties compared with small organic molecules [120,161].

It is well-documented that the addition of a small molecule drug to an otherwise soluble and stable antibody can cause aggregation and other physicochemical instability in the ADC [162,163]. This is not only because many of the small molecule drugs are bulky and hydrophobic in nature leading to a significant increase in the hydrophobicity of the ADC, but also because the conjugation can induce perturbations to secondary and tertiary structures of the antibody resulting in reduced conformational stabilities. To this point, a systematic study of trastuzumab, trastuzumab-MCC conjugate intermediate, and trastuzumab-DM1 found that both conjugates suffered decreased thermal stability and increased aggregation compared with trastuzumab [164]. Recently, the impact of drug conjugation on intra- and intermolecular interactions of trastuzumab-DM1 compared with trastuzumab was studied and the results confirmed that the lower colloidal stability and higher aggregation propensity for trastuzumab-DM1 are attributed to both reduced repulsive charge interaction and increased hydrophobicity [165]. Multiple publications have reported a more pronounced conjugation destabilizing effect on interchain cysteine conjugated ADCs and an inverse correlation between the drug load and stability [166–170]. Consistent with the findings on the interchain cysteine conjugated ADCs, the high DAR species in trastuzumab-DM1, a lysine conjugate, have also been found to be less stable and more prone to aggregation than the low DAR species [171]. Site-specific conjugation approaches with carefully chosen conjugation sites are expected to have less negative impact on stability and aggregation propensity of ADCs [64,65,172]. Most site-specific conjugates in the clinical pipeline have homogeneous DAR of 2 resulting in reduced hydrophobicity and aggregation compared to stochastic conjugates of average DAR of 3.5–4 where DAR species range from 0 to 8 or higher. In order to achieve sufficient efficacy with a relatively low DAR number, potent payloads such as PBD, a MDR1-resistant maytansine payload, and an auristatin payload Aur0101 have been developed. However, the DAR 2 site-specific conjugates with PBD and Aur0101 have shown limited therapeutic index in clinic thus far [173–176], while the clinical data for the DAR 2 maytansine ADC is pending [177]. While extensive characterization studies have been reported for antibody conjugated to cytotoxic payloads, there is a scarcity of literature on the molecular properties of conjugates with non-toxic small organic molecule payloads and nucleic acids. The optimal DAR number and solution property of conjugates with oligonucleotides, which are highly charged and significantly larger than small organic molecules, remains to be determined.

In addition to all the issues encountered during antibody formulation development, the formulation development of ADC drugs must find suitable pH and excipient conditions to simultaneously maintain the stability of the antibody, the linker, and the small molecule drug (reviewed in [162,163]). Even if there is an in-depth understanding of the stability of the parental antibody in aqueous solution, its physical stability may change upon conjugation in the presence of organic solvent or through possible cross-linking mechanism, and its chemical stability may depend on the conjugation method [178]. An example is the light-sensitivity in a model ADC using trastuzumab whereas trastuzumab itself does not show such sensitivity [179]. While much is known in the literature regarding the chemical stability of monoclonal antibodies [180] and the data on commonly used cytotoxic linker-payloads is accumulating [153,181,182], novel and non-toxic linker-payloads including siRNA will require a clear understanding of their degradation pathways in order to form control strategies during drug development process, similar to what has been demonstrated on payload metabolism [183]. Therefore, a comprehensive evaluation of the combined system will always be necessary for novel ADCs.

All the current ADC drugs on the market are lyophilized products suitable for intravenous administration. Such a freeze-dried state protects the ADCs from chemical degradation and aggregation which occur under long-term solution storage conditions. Although interest has been growing for liquid formulation based on the increased experience with ADCs and the improved solubility and stability of the new generation of ADCs, few such feasibility studies have been reported in the literature to date. It can be particularly challenging to prevent payloads from falling off the antibody over a long period of time in solution. In addition, the currently approved ADC drugs are reconstituted to 0.25–20 mg/mL in solution, significantly below the concentrations for most therapeutic antibody products. While the above is a viable approach for intravenous administration commonly used for oncology therapies, the emerging non-oncology application of the ADCs will likely demand stable liquid formulation and subcutaneous administration as commonly expected for many antibody drugs to increase convenience for patients. This growing trend exerts pressure on linker-payload design and conjugation methods in addition to the properties of the parental antibodies, as well as on formulation and device development. The current pre-clinical data for non-oncology ADCs suggest that the ADC doses might not be significantly lower than those for antibodies [41,45,47,52]. The clinical efficacy and therapeutic index of the non-oncology ADCs will ultimately determine the dose requirement and the appropriate drug product concentration.

8. Conclusions

Antibody conjugates in oncology have thus far delivered several successfully approved therapeutics. Extensive research into novel payloads, more developable linkers and conjugation chemistries further enable the field of oncology conjugates to cross the finish line. These learnings and advances also help propel the next generation of conjugates for non-oncology indications. Additional challenges such as in vivo stability, formulation and delivery will drive the field to seek solutions to broaden the therapeutic horizon to include payloads like nucleic acids. Overall, the rise of non-oncology ADC therapeutics offers a huge opportunity for innovation at multiple fronts of drug discovery and development for years to come.

Author Contributions: Writing—Original Draft Preparation, Review & Editing, D.L., J.M.W., T.L., R.M.M., A.D.-M. and Y.F. All authors have read and agreed to the published version of the manuscript.

Funding: This research received no external funding.

Conflicts of Interest: At the time this manuscript was prepared, all authors were employees of Eli Lilly and Company. All authors are recipients of Eli Lilly and Company stock grants. The authors declare no conflict of interest.

References

1. Singh, S.; Kumar, N.K.; Dwiwedi, P.; Charan, J.; Kaur, R.; Sidhu, P.; Chugh, V.K. Monoclonal antibodies: A review. *Curr. Clin. Pharmacol.* **2018**, *13*, 85–99. [CrossRef]
2. Grilo, A.L.; Mantalaris, A. The increasingly human and profitable monoclonal antibody market. *Trends Biotechnol.* **2019**, *37*, 9–16. [CrossRef]
3. Kaplon, H.; Reichert, J.M. Antibodies to watch in 2019. *mAbs* **2019**, *11*, 219–238. [CrossRef]
4. Sifniotis, V.; Cruz, E.; Eroglu, B.; Kayser, V. Current advancements in addressing key challenges of therapeutic antibody design, manufacture, and formulation. *Antibodies* **2019**, *8*, 36. [CrossRef]
5. Weiner, G.J. Building better monoclonal antibody-based therapeutics. *Nat. Rev. Cancer* **2015**, *15*, 361–370. [CrossRef] [PubMed]
6. Saeed, A.F.U.H.; Wang, R.; Ling, S.; Wang, S. Antibody engineering for pursuing a healthier future. *Front. Microbiol.* **2017**, *8*, 495. [CrossRef] [PubMed]
7. Kohler, G.; Milstein, C. Continuous cultures of fused cells secreting antibody of predefined specificity. *Nature* **1975**, *256*, 495–497. [CrossRef] [PubMed]
8. Riechmann, L.; Clark, M.; Waldmann, H.; Winter, G. Reshaping human antibodies for therapy. *Nature* **1988**, *332*, 323–327. [CrossRef]
9. Harding, F.A.; Stickler, M.M.; Razo, J.; DuBridge, R.B. The immunogenicity of humanized and fully human antibodies: Residual immunogenicity resides in the CDR regions. *mAbs* **2010**, *2*, 256–265. [CrossRef]
10. Strohl, W.R. Human antibody discovery platforms. In *Protein Therapeutics*; Wiley-VCH Verlag GmbH & Co. KgaA: Weinheim, Germany, 2017; pp. 113–159. [CrossRef]
11. Tsuchikama, K.; An, Z. Antibody-drug conjugates: Recent advances in conjugation and linker chemistries. *Protein Cell* **2018**, *9*, 33–46. [CrossRef]
12. Mukherjee, A.; Waters, A.K.; Babic, I.; Nurmemmedov, E.; Glassy, M.C.; Kesari, S.; Yenugonda, V.M. Antibody drug conjugates: Progress, pitfalls, and promises. *Hum. Antibodies* **2019**, *27*, 53–62. [CrossRef]
13. Beck, A.; Goetsch, L.; Dumontet, C.; Corvaia, N. Strategies and challenges for the next generation of antibody-drug conjugates. *Nat. Rev. Drug Discov.* **2017**, *16*, 315–337. [CrossRef] [PubMed]
14. Chau, C.H.; Steeg, P.S.; Figg, W.D. Antibody–drug conjugates for cancer. *Lancet* **2019**, *394*, 793–804. [CrossRef]
15. FDA Approves First Chemoimmunotherapy Regimen for Patients with Relapsed or Refractory Diffuse Large B-Cell Lymphoma. Available online: https://www.fda.gov/news-events/press-announcements/fda-approves-first-chemoimmunotherapy-regimen-patients-relapsed-or-refractory-diffuse-large-b-cell (accessed on 25 June 2019).
16. Petersen, B.H.; DeHerdt, S.V.; Schneck, D.W.; Bumol, T.F. The human immune response to KS1/4-desacetylvinblastine (LY256787) and KS1/4-desacetylvinblastine hydrazide (LY203728) in single and multiple dose clinical studies. *Cancer Res.* **1991**, *51*, 2286–2290. [PubMed]
17. Abdollahpour-Alitappeh, M.; Lotfinia, M.; Gharibi, T.; Mardaneh, J.; Farhadihosseinabadi, B.; Larki, P.; Faghfourian, B.; Sepehr, K.S.; Abbaszadeh-Goudarzi, K.; Abbaszadeh-Goudarzi, G.; et al. Antibody-drug conjugates (ADCs) for cancer therapy: Strategies, challenges, and successes. *J. Cell. Physiol.* **2019**, *234*, 5628–5642. [CrossRef] [PubMed]
18. Boehncke, W.H.; Brembilla, N.C. Immunogenicity of biologic therapies: Causes and consequences. *Expert Rev. Clin. Immunol.* **2018**, *14*, 513–523. [CrossRef] [PubMed]

19. Pineda, C.; Castaneda Hernandez, G.; Jacobs, I.A.; Alvarez, D.F.; Carini, C. Assessing the immunogenicity of biopharmaceuticals. *BioDrugs* **2016**, *30*, 195–206. [CrossRef]
20. Ghetie, V.; Vitetta, E. Immunotoxins in the therapy of cancer: From bench to clinic. *Pharmacol. Ther.* **1994**, *63*, 209–234. [CrossRef]
21. Dumontet, C.; Jordan, M.A. Microtubule-binding agents: A dynamic field of cancer therapeutics. *Nat. Rev. Drug Discov.* **2010**, *9*, 790–803. [CrossRef]
22. Pyzik, M.; Sand, K.M.K.; Hubbard, J.J.; Andersen, J.T.; Sandlie, I.; Blumberg, R.S. The neonatal Fc Receptor (FcRn): A misnomer? *Front. Immunol.* **2019**, *10*, 1540. [CrossRef]
23. Datta-Mannan, A.; Choi, H.; Stokell, D.; Tang, J.; Murphy, A.; Wrobleski, A.; Feng, Y. The properties of cysteine-conjugated antibody-drug conjugates are impacted by the IgG subclass. *AAPS J.* **2018**, *20*, 103. [CrossRef]
24. Mohammed, R.; Milne, A.; Kayani, K.; Ojha, U. How the discovery of rituximab impacted the treatment of B-cell non-Hodgkin's lymphomas. *J. Blood Med.* **2019**, *10*, 71–84. [CrossRef] [PubMed]
25. Hoffmann, R.M.; Coumbe, B.G.T.; Josephs, D.H.; Mele, S.; Ilieva, K.M.; Cheung, A.; Tutt, A.N.; Spicer, J.F.; Thurston, D.E.; Crescioli, S.; et al. Antibody structure and engineering considerations for the design and function of Antibody Drug Conjugates (ADCs). *Oncoimmunology* **2018**, *7*, e1395127. [CrossRef] [PubMed]
26. Junttila, T.T.; Li, G.; Parsons, K.; Phillips, G.L.; Sliwkowski, M.X. Trastuzumab-DM1 (T-DM1) retains all the mechanisms of action of trastuzumab and efficiently inhibits growth of lapatinib insensitive breast cancer. *Breast Cancer Res. Treat.* **2011**, *128*, 347–356. [CrossRef] [PubMed]
27. English, D.P.; Bellone, S.; Schwab, C.L.; Bortolomai, I.; Bonazzoli, E.; Cocco, E.; Buza, N.; Hui, P.; Lopez, S.; Ratner, E.; et al. T-DM1, a novel antibody-drug conjugate, is highly effective against primary HER2 overexpressing uterine serous carcinoma in vitro and in vivo. *Cancer Med.* **2014**, *3*, 1256–1265. [CrossRef] [PubMed]
28. Nicoletti, R.; Lopez, S.; Bellone, S.; Cocco, E.; Schwab, C.L.; Black, J.D.; Centritto, F.; Zhu, L.; Bonazzoli, E.; Buza, N.; et al. T-DM1, a novel antibody-drug conjugate, is highly effective against uterine and ovarian carcinosarcomas overexpressing HER2. *Clin. Exp. Metastasis* **2015**, *32*, 29–38. [CrossRef] [PubMed]
29. Black, J.; Menderes, G.; Bellone, S.; Schwab, C.L.; Bonazzoli, E.; Ferrari, F.; Predolini, F.; De Haydu, C.; Cocco, E.; Buza, N.; et al. SYD985, a novel duocarmycin-based HER2-targeting antibody-drug conjugate, shows antitumor activity in uterine serous carcinoma with HER2/Neu expression. *Mol. Cancer Ther.* **2016**, *15*, 1900–1909. [CrossRef]
30. Menderes, G.; Bonazzoli, E.; Bellone, S.; Black, J.; Altwerger, G.; Masserdotti, A.; Pettinella, F.; Zammataro, L.; Buza, N.; Hui, P.; et al. SYD985, a novel duocarmycin-based HER2-targeting antibody-drug conjugate, shows promising antitumor activity in epithelial ovarian carcinoma with HER2/Neu expression. *Gynecol. Oncol.* **2017**, *146*, 179–186. [CrossRef]
31. Ogitani, Y.; Aida, T.; Hagihara, K.; Yamaguchi, J.; Ishii, C.; Harada, N.; Soma, M.; Okamoto, H.; Oitate, M.; Arakawa, S.; et al. DS-8201a, a novel HER2-targeting ADC with a novel DNA topoisomerase i inhibitor, demonstrates a promising antitumor efficacy with differentiation from T-DM1. *Clin. Cancer Res.* **2016**, *22*, 5097–5108. [CrossRef]
32. Cardillo, T.M.; Govindan, S.V.; Sharkey, R.M.; Trisal, P.; Arrojo, R.; Liu, D.; Rossi, E.A.; Chang, C.H.; Goldenberg, D.M. Sacituzumab govitecan (IMMU-132), an Anti-Trop-2/SN-38 antibody-drug conjugate: Characterization and efficacy in pancreatic, gastric, and other cancers. *Bioconjug. Chem.* **2015**, *26*, 919–931. [CrossRef]
33. Bardia, A.; Mayer, I.A.; Vahdat, L.T.; Tolaney, S.M.; Isakoff, S.J.; Diamond, J.R.; O'Shaughnessy, J.; Moroose, R.L.; Santin, A.D.; Abramson, V.G.; et al. Sacituzumab govitecan-hziy in refractory metastatic triple-negative breast cancer. *N. Engl. J. Med.* **2019**, *380*, 741–751. [CrossRef] [PubMed]
34. Heist, R.S.; Guarino, M.J.; Masters, G.; Purcell, W.T.; Starodub, A.N.; Horn, L.; Scheff, R.J.; Bardia, A.; Messersmith, W.A.; Berlin, J.; et al. Therapy of advanced non-small-cell lung cancer with an SN-38-Anti-Trop-2 drug conjugate, sacituzumab govitecan. *J. Clin. Oncol.* **2017**, *35*, 2790–2797. [CrossRef] [PubMed]
35. Pereira, N.A.; Chan, K.F.; Lin, P.C.; Song, Z. The "less-is-more" in therapeutic antibodies: Afucosylated anti-cancer antibodies with enhanced antibody-dependent cellular cytotoxicity. *mAbs* **2018**, *10*, 693–711. [CrossRef] [PubMed]

36. Tai, Y.-T.; Mayes, P.A.; Acharya, C.; Zhong, M.Y.; Cea, M.; Cagnetta, A.; Craigen, J.; Yates, J.; Gliddon, L.; Fieles, W.; et al. Novel anti-B-cell maturation antigen antibody-drug conjugate (GSK2857916) selectively induces killing of multiple myeloma. *Blood* **2014**, *123*, 3128–3138. [CrossRef] [PubMed]
37. Trudel, S.; Lendvai, N.; Popat, R.; Voorhees, P.M.; Reeves, B.; Libby, E.N.; Richardson, P.G.; Hoos, A.; Gupta, I.; Bragulat, V.; et al. Antibody–drug conjugate, GSK2857916, in relapsed/refractory multiple myeloma: An update on safety and efficacy from dose expansion phase I study. *Blood Cancer J.* **2019**, *9*, 37. [CrossRef]
38. Szot, C.; Saha, S.; Zhang, X.M.; Zhu, Z.; Hilton, M.B.; Morris, K.; Seaman, S.; Dunleavey, J.M.; Hsu, K.-S.; Yu, G.-J.; et al. Tumor stroma–targeted antibody-drug conjugate triggers localized anticancer drug release. *J. Clin. Investig.* **2018**, *128*, 2927–2943. [CrossRef]
39. Liu, R.; Wang, R.E.; Wang, F. Antibody-drug conjugates for non-oncological indications. *Expert Opin. Biol. Ther.* **2016**, *16*, 591–593. [CrossRef]
40. Yu, S.; Lim, A.; Tremblay, M.S. Next horizons: ADCs beyond oncology. In *Innovations for Next-Generation Antibody-Drug Conjugates*; Humana Press: Totowa, NJ, USA, 2018; pp. 321–347. [CrossRef]
41. Lehar, S.M.; Pillow, T.; Xu, M.; Staben, L.; Kajihara, K.K.; Vandlen, R.; DePalatis, L.; Raab, H.; Hazenbos, W.L.; Morisaki, J.H.; et al. Novel antibody-antibiotic conjugate eliminates intracellular S. aureus. *Nature* **2015**, *527*, 323–328. [CrossRef]
42. Mariathasan, S.; Tan, M.-W. Antibody–antibiotic conjugates: A novel therapeutic platform against bacterial infections. *Trends Mol. Med.* **2017**, *23*, 135–149. [CrossRef]
43. Everts, M.; Kok, R.J.; Ásgeirsdóttir, S.A.; Melgert, B.N.; Moolenaar, T.J.M.; Koning, G.A.; van Luyn, M.J.A.; Meijer, D.K.F.; Molema, G. Selective intracellular delivery of dexamethasone into activated endothelial cells using an e-selectin-directed immunoconjugate. *J. Immunol.* **2002**, *168*, 883–889. [CrossRef]
44. Lim, R.K.; Yu, S.; Cheng, B.; Li, S.; Kim, N.J.; Cao, Y.; Chi, V.; Kim, J.Y.; Chatterjee, A.K.; Schultz, P.G.; et al. Targeted delivery of LXR agonist using a site-specific antibody-drug conjugate. *Bioconjug. Chem.* **2015**, *26*, 2216–2222. [CrossRef] [PubMed]
45. Yu, S.; Pearson, A.D.; Lim, R.K.; Rodgers, D.T.; Li, S.; Parker, H.B.; Weglarz, M.; Hampton, E.N.; Bollong, M.J.; Shen, J.; et al. Targeted delivery of an anti-inflammatory PDE4 inhibitor to immune cells via an antibody-drug conjugate. *Mol. Ther.* **2016**, *24*, 2078–2089. [CrossRef]
46. Beaumont, M.; Tomazela, D.; Hodges, D.; Ermakov, G.; Hsieh, E.; Figueroa, I.; So, O.-Y.; Song, Y.; Ma, H.; Antonenko, S.; et al. Antibody-drug conjugates: Integrated bioanalytical and biodisposition assessments in lead optimization and selection. *AAPS Open* **2018**, *4*, 6. [CrossRef]
47. Kvirkvelia, N.; McMenamin, M.; Gutierrez, V.I.; Lasareishvili, B.; Madaio, M.P. Human anti-alpha3(IV)NC1 antibody drug conjugates target glomeruli to resolve nephritis. *Am. J. Physiol. Ren. Physiol.* **2015**, *309*, F680–F684. [CrossRef] [PubMed]
48. Wang, R.E.; Liu, T.; Wang, Y.; Cao, Y.; Du, J.; Luo, X.; Deshmukh, V.; Kim, C.H.; Lawson, B.R.; Tremblay, M.S.; et al. An immunosuppressive antibody-drug conjugate. *J. Am. Chem. Soc.* **2015**, *137*, 3229–3232. [CrossRef] [PubMed]
49. Palchaudhuri, R.; Saez, B.; Hoggatt, J.; Schajnovitz, A.; Sykes, D.B.; Tate, T.A.; Czechowicz, A.; Kfoury, Y.; Ruchika, F.; Rossi, D.J.; et al. Non-genotoxic conditioning for hematopoietic stem cell transplantation using a hematopoietic-cell-specific internalizing immunotoxin. *Nat. Biotechnol.* **2016**, *34*, 738–745. [CrossRef]
50. Kern, J.C.; Dooney, D.; Zhang, R.; Liang, L.; Brandish, P.E.; Cheng, M.; Feng, G.; Beck, A.; Bresson, D.; Firdos, J.; et al. Novel phosphate modified cathepsin B linkers: Improving aqueous solubility and enhancing payload scope of ADCs. *Bioconjug. Chem.* **2016**, *27*, 2081–2088. [CrossRef]
51. Brandish, P.E.; Palmieri, A.; Antonenko, S.; Beaumont, M.; Benso, L.; Cancilla, M.; Cheng, M.; Fayadat-Dilman, L.; Feng, G.; Figueroa, I.; et al. Development of Anti-CD74 antibody-drug conjugates to target glucocorticoids to immune cells. *Bioconjug. Chem.* **2018**, *29*, 2357–2369. [CrossRef]
52. Graversen, J.H.; Svendsen, P.; Dagnaes-Hansen, F.; Dal, J.; Anton, G.; Etzerodt, A.; Petersen, M.D.; Christensen, P.A.; Moller, H.J.; Moestrup, S.K. Targeting the hemoglobin scavenger receptor CD163 in macrophages highly increases the anti-inflammatory potency of dexamethasone. *Mol. Ther.* **2012**, *20*, 1550–1558. [CrossRef]
53. Thomsen, K.L.; Møller, H.J.; Graversen, J.H.; Magnusson, N.E.; Moestrup, S.K.; Vilstrup, H.; Grønbæk, H. Anti-CD163-dexamethasone conjugate inhibits the acute phase response to lipopolysaccharide in rats. *World J. Hepatol.* **2016**, *8*, 726–730. [CrossRef]

54. Yarian, F.; Alibakhshi, A.; Eyvazi, S.; Arezumand, R.; Ahangarzadeh, S. Antibody-drug therapeutic conjugates: Potential of antibody-siRNAs in cancer therapy. *J. Cell. Physiol.* **2019**. [CrossRef] [PubMed]
55. Lu, H.; Wang, D.; Kazane, S.; Javahishvili, T.; Tian, F.; Song, F.; Sellers, A.; Barnett, B.; Schultz, P.G. Site-specific antibody–polymer conjugates for siRNA delivery. *J. Am. Chem. Soc.* **2013**, *135*, 13885–13891. [CrossRef] [PubMed]
56. Zhou, Q. Site-specific antibody conjugation for ADC and beyond. *Biomedicines* **2017**, *5*, 64. [CrossRef] [PubMed]
57. Behrens, C.R.; Liu, B. Methods for site-specific drug conjugation to antibodies. *mAbs* **2014**, *6*, 46–53. [CrossRef] [PubMed]
58. Schumacher, D.; Hackenberger, C.P.; Leonhardt, H.; Helma, J. Current status: Site-specific antibody drug conjugates. *J. Clin. Immunol.* **2016**, *36* (Suppl. S1), 100–107. [CrossRef] [PubMed]
59. Hamblett, K.J.; Senter, P.D.; Chace, D.F.; Sun, M.M.; Lenox, J.; Cerveny, C.G.; Kissler, K.M.; Bernhardt, S.X.; Kopcha, A.K.; Zabinski, R.F.; et al. Effects of drug loading on the antitumor activity of a monoclonal antibody drug conjugate. *Clin. Cancer Res.* **2004**, *10*, 7063–7070. [CrossRef]
60. Junutula, J.R.; Flagella, K.M.; Graham, R.A.; Parsons, K.L.; Ha, E.; Raab, H.; Bhakta, S.; Nguyen, T.; Dugger, D.L.; Li, G.; et al. Engineered thio-trastuzumab-DM1 conjugate with an improved therapeutic index to target human epidermal growth factor receptor 2-positive breast cancer. *Clin. Cancer Res.* **2010**, *16*, 4769–4778. [CrossRef]
61. Junutula, J.R.; Raab, H.; Clark, S.; Bhakta, S.; Leipold, D.D.; Weir, S.; Chen, Y.; Simpson, M.; Tsai, S.P.; Dennis, M.S.; et al. Site-specific conjugation of a cytotoxic drug to an antibody improves the therapeutic index. *Nat. Biotechnol.* **2008**, *26*, 925–932. [CrossRef]
62. Shen, B.Q.; Xu, K.; Liu, L.; Raab, H.; Bhakta, S.; Kenrick, M.; Parsons-Reponte, K.L.; Tien, J.; Yu, S.F.; Mai, E.; et al. Conjugation site modulates the in vivo stability and therapeutic activity of antibody-drug conjugates. *Nat. Biotechnol.* **2012**, *30*, 184–189. [CrossRef]
63. Dimasi, N.; Fleming, R.; Zhong, H.; Bezabeh, B.; Kinneer, K.; Christie, R.J.; Fazenbaker, C.; Wu, H.; Gao, C. Efficient preparation of site-specific antibody-drug conjugates using cysteine insertion. *Mol. Pharm.* **2017**, *14*, 1501–1516. [CrossRef]
64. Sussman, D.; Westendorf, L.; Meyer, D.W.; Leiske, C.I.; Anderson, M.; Okeley, N.M.; Alley, S.C.; Lyon, R.; Sanderson, R.J.; Carter, P.J.; et al. Engineered cysteine antibodies: An improved antibody-drug conjugate platform with a novel mechanism of drug-linker stability. *Protein Eng. Des. Sel.* **2018**, *31*, 47–54. [CrossRef] [PubMed]
65. Strop, P.; Liu, S.H.; Dorywalska, M.; Delaria, K.; Dushin, R.G.; Tran, T.T.; Ho, W.H.; Farias, S.; Casas, M.G.; Abdiche, Y.; et al. Location matters: Site of conjugation modulates stability and pharmacokinetics of antibody drug conjugates. *Chem. Biol.* **2013**, *20*, 161–167. [CrossRef] [PubMed]
66. Falck, G.; Müller, K.M. Enzyme-based labeling strategies for antibody–drug conjugates and antibody mimetics. *Antibodies* **2018**, *7*, 4. [CrossRef] [PubMed]
67. Dennler, P.; Chiotellis, A.; Fischer, E.; Bregeon, D.; Belmant, C.; Gauthier, L.; Lhospice, F.; Romagne, F.; Schibli, R. Transglutaminase-based chemo-enzymatic conjugation approach yields homogeneous antibody-drug conjugates. *Bioconjug. Chem.* **2014**, *25*, 569–578. [CrossRef] [PubMed]
68. Jeger, S.; Zimmermann, K.; Blanc, A.; Grunberg, J.; Honer, M.; Hunziker, P.; Struthers, H.; Schibli, R. Site-specific and stoichiometric modification of antibodies by bacterial transglutaminase. *Angew. Chem. Int. Ed. Engl.* **2010**, *49*, 9995–9997. [CrossRef] [PubMed]
69. Dorywalska, M.; Strop, P.; Melton-Witt, J.A.; Hasa-Moreno, A.; Farias, S.E.; Galindo Casas, M.; Delaria, K.; Lui, V.; Poulsen, K.; Loo, C.; et al. Effect of attachment site on stability of cleavable antibody drug conjugates. *Bioconjug. Chem.* **2015**, *26*, 650–659. [CrossRef]
70. Ritzefeld, M. Sortagging: A robust and efficient chemoenzymatic ligation strategy. *Chemistry* **2014**, *20*, 8516–8529. [CrossRef]
71. Beerli, R.R.; Hell, T.; Merkel, A.S.; Grawunder, U. Sortase enzyme-mediated generation of site-specifically conjugated antibody drug conjugates with high in vitro and in vivo potency. *PLoS ONE* **2015**, *10*, e0131177. [CrossRef]
72. Pishesha, N.; Ingram, J.R.; Ploegh, H.L. Sortase A: A model for transpeptidation and its biological applications. *Annu. Rev. Cell Dev. Biol.* **2018**, *34*, 163–188. [CrossRef]

73. Carrico, I.S.; Carlson, B.L.; Bertozzi, C.R. Introducing genetically encoded aldehydes into proteins. *Nat. Chem. Biol.* **2007**, *3*, 321. [CrossRef]
74. Rabuka, D.; Rush, J.S.; deHart, G.W.; Wu, P.; Bertozzi, C.R. Site-specific chemical protein conjugation using genetically encoded aldehyde tags. *Nat. Protoc.* **2012**, *7*, 1052–1067. [CrossRef] [PubMed]
75. Liu, J.; Barfield, R.M.; Rabuka, D. Site-specific bioconjugation using SMARTag((R)) Technology: A practical and effective chemoenzymatic approach to generate antibody-drug conjugates. *Methods Mol. Biol.* **2019**, *2033*, 131–147. [CrossRef] [PubMed]
76. Agarwal, P.; van der Weijden, J.; Sletten, E.M.; Rabuka, D.; Bertozzi, C.R. A Pictet-Spengler ligation for protein chemical modification. *Proc. Natl. Acad. Sci. USA* **2013**, *110*, 46–51. [CrossRef] [PubMed]
77. Agarwal, P.; Bertozzi, C.R. Site-specific antibody–drug conjugates: The nexus of bioorthogonal chemistry, protein engineering, and drug development. *Bioconjug. Chem.* **2015**, *26*, 176–192. [CrossRef] [PubMed]
78. Hallam, T.J.; Smider, V.V. Unnatural amino acids in novel antibody conjugates. *Future Med. Chem.* **2014**, *6*, 1309–1324. [CrossRef] [PubMed]
79. Kim, C.H.; Axup, J.Y.; Schultz, P.G. Protein conjugation with genetically encoded unnatural amino acids. *Curr. Opin. Chem. Biol.* **2013**, *17*, 412–419. [CrossRef]
80. Axup, J.Y.; Bajjuri, K.M.; Ritland, M.; Hutchins, B.M.; Kim, C.H.; Kazane, S.A.; Halder, R.; Forsyth, J.S.; Santidrian, A.F.; Stafin, K.; et al. Synthesis of site-specific antibody-drug conjugates using unnatural amino acids. *Proc. Natl. Acad. Sci. USA* **2012**, *109*, 16101–16106. [CrossRef]
81. Tian, F.; Lu, Y.; Manibusan, A.; Sellers, A.; Tran, H.; Sun, Y.; Phuong, T.; Barnett, R.; Hehli, B.; Song, F.; et al. A general approach to site-specific antibody drug conjugates. *Proc. Natl. Acad. Sci. USA* **2014**, *111*, 1766–1771. [CrossRef]
82. Hofer, T.; Skeffington, L.R.; Chapman, C.M.; Rader, C. Molecularly defined antibody conjugation through a selenocysteine interface. *Biochemistry* **2009**, *48*, 12047–12057. [CrossRef]
83. Zimmerman, E.S.; Heibeck, T.H.; Gill, A.; Li, X.; Murray, C.J.; Madlansacay, M.R.; Tran, C.; Uter, N.T.; Yin, G.; Rivers, P.J.; et al. Production of site-specific antibody–drug conjugates using optimized non-natural amino acids in a cell-free expression system. *Bioconjug. Chem.* **2014**, *25*, 351–361. [CrossRef]
84. Yamada, K.; Ito, Y. Recent chemical approaches for site-specific conjugation of native antibodies: Technologies toward next generation antibody-drug conjugates. *ChemBioChem* **2019**, *20*, 2729–2737. [CrossRef] [PubMed]
85. Ban, H.; Nagano, M.; Gavrilyuk, J.; Hakamata, W.; Inokuma, T.; Barbas, C.F., 3rd. Facile and stabile linkages through tyrosine: Bioconjugation strategies with the tyrosine-click reaction. *Bioconjug. Chem.* **2013**, *24*, 520–532. [CrossRef] [PubMed]
86. Zuberbuhler, K.; Casi, G.; Bernardes, G.J.; Neri, D. Fucose-specific conjugation of hydrazide derivatives to a vascular-targeting monoclonal antibody in IgG format. *Chem. Commun.* **2012**, *48*, 7100–7102. [CrossRef] [PubMed]
87. Zhou, Q.; Stefano, J.E.; Manning, C.; Kyazike, J.; Chen, B.; Gianolio, D.A.; Park, A.; Busch, M.; Bird, J.; Zheng, X.; et al. Site-specific antibody–drug conjugation through glycoengineering. *Bioconjug. Chem.* **2014**, *25*, 510–520. [CrossRef] [PubMed]
88. van Geel, R.; Wijdeven, M.A.; Heesbeen, R.; Verkade, J.M.; Wasiel, A.A.; van Berkel, S.S.; van Delft, F.L. Chemoenzymatic conjugation of toxic payloads to the globally conserved N-Glycan of native mAbs provides homogeneous and highly efficacious antibody-drug conjugates. *Bioconjug. Chem.* **2015**, *26*, 2233–2242. [CrossRef] [PubMed]
89. Li, X.; Fang, T.; Boons, G.J. Preparation of well-defined antibody-drug conjugates through glycan remodeling and strain-promoted azide-alkyne cycloadditions. *Angew. Chem. Int. Ed. Engl.* **2014**, *53*, 7179–7182. [CrossRef]
90. Badescu, G.; Bryant, P.; Bird, M.; Henseleit, K.; Swierkosz, J.; Parekh, V.; Tommasi, R.; Pawlisz, E.; Jurlewicz, K.; Farys, M.; et al. Bridging disulfides for stable and defined antibody drug conjugates. *Bioconjug. Chem.* **2014**, *25*, 1124–1136. [CrossRef]
91. Behrens, C.R.; Ha, E.H.; Chinn, L.L.; Bowers, S.; Probst, G.; Fitch-Bruhns, M.; Monteon, J.; Valdiosera, A.; Bermudez, A.; Liao-Chan, S.; et al. Antibody-Drug Conjugates (ADCs) derived from interchain cysteine cross-linking demonstrate improved homogeneity and other pharmacological properties over conventional heterogeneous ADCs. *Mol. Pharm.* **2015**, *12*, 3986–3998. [CrossRef]

92. Bryant, P.; Pabst, M.; Badescu, G.; Bird, M.; McDowell, W.; Jamieson, E.; Swierkosz, J.; Jurlewicz, K.; Tommasi, R.; Henseleit, K.; et al. In vitro and in vivo evaluation of cysteine rebridged trastuzumab-MMAE antibody drug conjugates with defined drug-to-antibody ratios. *Mol. Pharm.* **2015**, *12*, 1872–1879. [CrossRef]
93. Forte, N.; Chudasama, V.; Baker, J.R. Homogeneous antibody-drug conjugates via site-selective disulfide bridging. *Drug Discov. Today Technol.* **2018**, *30*, 11–20. [CrossRef]
94. Schumacher, F.F.; Nunes, J.P.M.; Maruani, A.; Chudasama, V.; Smith, M.E.B.; Chester, K.A.; Baker, J.R.; Caddick, S. Next generation maleimides enable the controlled assembly of antibody-drug conjugates via native disulfide bond bridging. *Org. Biomol. Chem.* **2014**, *12*, 7261–7269. [CrossRef] [PubMed]
95. Altwerger, G.; Bonazzoli, E.; Bellone, S.; Egawa-Takata, T.; Menderes, G.; Pettinella, F.; Bianchi, A.; Riccio, F.; Feinberg, J.; Zammataro, L.; et al. In vitro and in vivo activity of IMGN853, an antibody-drug conjugate targeting folate receptor alpha linked to DM4, in biologically aggressive endometrial cancers. *Mol. Cancer Ther.* **2018**, *17*, 1003–1011. [CrossRef] [PubMed]
96. Pei, Z.; Chen, C.; Chen, J.; Cruz-Chuh, J.D.; Delarosa, R.; Deng, Y.; Fourie-O'Donohue, A.; Figueroa, I.; Guo, J.; Jin, W.; et al. Exploration of pyrrolobenzodiazepine (PBD)-dimers containing disulfide-based prodrugs as payloads for antibody-drug conjugates. *Mol. Pharm.* **2018**, *15*, 3979–3996. [CrossRef] [PubMed]
97. Mantaj, J.; Jackson, P.J.M.; Rahman, K.M.; Thurston, D.E. From anthramycin to pyrrolobenzodiazepine (PBD)-containing antibody-drug conjugates (ADCs). *Angew. Chem. Int. Ed. Engl.* **2017**, *56*, 462–488. [CrossRef]
98. Dan, N.; Setua, S.; Kashyap, V.K.; Khan, S.; Jaggi, M.; Yallapu, M.M.; Chauhan, S.C. Antibody-drug conjugates for cancer therapy: Chemistry to clinical implications. *Pharmaceuticals* **2018**, *11*, 32. [CrossRef]
99. Ratnayake, A.S.; Chang, L.P.; Tumey, L.N.; Loganzo, F.; Chemler, J.A.; Wagenaar, M.; Musto, S.; Li, F.; Janso, J.E.; Ballard, T.E.; et al. Natural product bis-intercalator depsipeptides as a new class of payloads for antibody-drug conjugates. *Bioconjug. Chem.* **2019**, *30*, 200–209. [CrossRef]
100. Lerchen, H.G.; Wittrock, S.; Stelte-Ludwig, B.; Sommer, A.; Berndt, S.; Griebenow, N.; Rebstock, A.S.; Johannes, S.; Cancho-Grande, Y.; Mahlert, C.; et al. Antibody-drug conjugates with pyrrole-based KSP inhibitors as the payload class. *Angew. Chem. Int. Ed. Engl.* **2018**, *57*, 15243–15247. [CrossRef]
101. Karpov, A.S.; Abrams, T.; Clark, S.; Raikar, A.; D'Alessio, J.A.; Dillon, M.P.; Gesner, T.G.; Jones, D.; Lacaud, M.; Mallet, W.; et al. Nicotinamide phosphoribosyltransferase inhibitor as a novel payload for antibody-drug conjugates. *ACS Med. Chem. Lett.* **2018**, *9*, 838–842. [CrossRef]
102. Love, E.A.; Sattikar, A.; Cook, H.; Gillen, K.; Large, J.M.; Patel, S.; Matthews, D.; Merritt, A. Developing an antibody–drug conjugate approach to selective inhibition of an extracellular protein. *ChemBioChem* **2019**, *20*, 754–758. [CrossRef]
103. Brown, E.D.; Wright, G.D. Antibacterial drug discovery in the resistance era. *Nature* **2016**, *529*, 336–343. [CrossRef]
104. van der Goes, M.C.; Jacobs, J.W.; Bijlsma, J.W. The value of glucocorticoid co-therapy in different rheumatic diseases-positive and adverse effects. *Arthritis Res. Ther.* **2014**, *16*, S2. [CrossRef] [PubMed]
105. Schäcke, H.; Döcke, W.-D.; Asadullah, K. Mechanisms involved in the side effects of glucocorticoids. *Pharmacol. Ther.* **2002**, *96*, 23–43. [CrossRef]
106. Kern, J.C.; Cancilla, M.; Dooney, D.; Kwasnjuk, K.; Zhang, R.; Beaumont, M.; Figueroa, I.; Hsieh, S.; Liang, L.; Tomazela, D.; et al. Discovery of pyrophosphate diesters as tunable, soluble, and bioorthogonal linkers for site-specific antibody-drug conjugates. *J. Am. Chem. Soc.* **2016**, *138*, 1430–1445. [CrossRef] [PubMed]
107. Peters, C.; Brown, S. Antibody–drug conjugates as novel anti-cancer chemotherapeutics. *Biosci. Rep.* **2015**, *35*, e00225. [CrossRef] [PubMed]
108. Juliano, R.L. The delivery of therapeutic oligonucleotides. *Nucleic Acids Res.* **2016**, *44*, 6518–6548. [CrossRef]
109. Dovgan, I.; Koniev, O.; Kolodych, S.; Wagner, A. Antibody–oligonucleotide conjugates as therapeutic, imaging, and detection agents. *Bioconjug. Chem.* **2019**, *30*, 2483–2501. [CrossRef]
110. Levin, A.A. Treating disease at the RNA level with oligonucleotides. *N. Engl. J. Med.* **2019**, *380*, 57–70. [CrossRef]
111. Geary, R.S.; Norris, D.; Yu, R.; Bennett, C.F. Pharmacokinetics, biodistribution and cell uptake of antisense oligonucleotides. *Adv. Drug Deliv. Rev.* **2015**, *87*, 46–51. [CrossRef]
112. Park, J.; Park, J.; Pei, Y.; Xu, J.; Yeo, Y. Pharmacokinetics and biodistribution of recently-developed siRNA nanomedicines. *Adv. Drug Deliv. Rev.* **2016**, *104*, 93–109. [CrossRef]

113. Benizri, S.; Gissot, A.; Martin, A.; Vialet, B.; Grinstaff, M.W.; Barthélémy, P. Bioconjugated oligonucleotides: Recent developments and therapeutic applications. *Bioconjug. Chem.* **2019**, *30*, 366–383. [CrossRef]
114. Nair, J.K.; Willoughby, J.L.S.; Chan, A.; Charisse, K.; Alam, M.R.; Wang, Q.; Hoekstra, M.; Kandasamy, P.; Kel'in, A.V.; Milstein, S.; et al. Multivalent N-acetylgalactosamine-conjugated siRNA localizes in hepatocytes and elicits robust RNAi-mediated gene silencing. *J. Am. Chem. Soc.* **2014**, *136*, 16958–16961. [CrossRef]
115. Sewing, S.; Gubler, M.; Gérard, R.; Avignon, B.; Mueller, Y.; Braendli-Baiocco, A.; Odin, M.; Moisan, A. GalNAc conjugation attenuates the cytotoxicity of antisense oligonucleotide drugs in renal tubular cells. *Mol. Ther. Nucleic Acids* **2019**, *14*, 67–79. [CrossRef] [PubMed]
116. Ämmälä, C.; Drury, W.J.; Knerr, L.; Ahlstedt, I.; Stillemark-Billton, P.; Wennberg-Huldt, C.; Andersson, E.-M.; Valeur, E.; Jansson-Löfmark, R.; Janzén, D.; et al. Targeted delivery of antisense oligonucleotides to pancreatic β-cells. *Sci. Adv.* **2018**, *4*, eaat3386. [CrossRef]
117. Osborn, M.F.; Coles, A.H.; Biscans, A.; Haraszti, R.A.; Roux, L.; Davis, S.; Ly, S.; Echeverria, D.; Hassler, M.R.; Godinho, B.M.D.C.; et al. Hydrophobicity drives the systemic distribution of lipid-conjugated siRNAs via lipid transport pathways. *Nucleic Acids Res.* **2018**, *47*, 1070–1081. [CrossRef] [PubMed]
118. Ma, Y.; Kowolik, C.M.; Swiderski, P.M.; Kortylewski, M.; Yu, H.; Horne, D.A.; Jove, R.; Caballero, O.L.; Simpson, A.J.G.; Lee, F.-T.; et al. Humanized lewis-y specific antibody based delivery of STAT3 siRNA. *ACS Chem. Biol.* **2011**, *6*, 962–970. [CrossRef] [PubMed]
119. Cuellar, T.L.; Barnes, D.; Nelson, C.; Tanguay, J.; Yu, S.-F.; Wen, X.; Scales, S.J.; Gesch, J.; Davis, D.; van Brabant Smith, A.; et al. Systematic evaluation of antibody-mediated siRNA delivery using an industrial platform of THIOMAB–siRNA conjugates. *Nucleic Acids Res.* **2014**, *43*, 1189–1203. [CrossRef]
120. Sugo, T.; Terada, M.; Oikawa, T.; Miyata, K.; Nishimura, S.; Kenjo, E.; Ogasawara-Shimizu, M.; Makita, Y.; Imaichi, S.; Murata, S.; et al. Development of antibody-siRNA conjugate targeted to cardiac and skeletal muscles. *J. Control. Release* **2016**, *237*, 1–13. [CrossRef]
121. Satake, N.; Duong, C.; Yoshida, S.; Oestergaard, M.; Chen, C.; Peralta, R.; Guo, S.; Seth, P.P.; Li, Y.; Beckett, L.; et al. Novel targeted therapy for precursor B cell acute lymphoblastic leukemia: Anti-CD22 antibody-MXD3 antisense oligonucleotide conjugate. *Mol. Med.* **2016**, *22*, 632–642. [CrossRef]
122. Arnold, A.E.; Malek-Adamian, E.; Le, P.U.; Meng, A.; Martinez-Montero, S.; Petrecca, K.; Damha, M.J.; Shoichet, M.S. Antibody-antisense oligonucleotide conjugate downregulates a key gene in glioblastoma stem cells. *Mol. Ther. Nucleic Acids* **2018**, *11*, 518–527. [CrossRef]
123. Humphreys, S.C.; Thayer, M.B.; Campuzano, I.D.G.; Netirojjanakul, C.; Rock, B.M. Quantification of siRNA-antibody conjugates in biological matrices by triplex-forming oligonucleotide ELISA. *Nucleic Acid Ther.* **2019**, *29*, 161–166. [CrossRef]
124. Bargh, J.D.; Isidro-Llobet, A.; Parker, J.S.; Spring, D.R. Cleavable linkers in antibody–drug conjugates. *Chem. Soc. Rev.* **2019**, *48*, 4361–4374. [CrossRef] [PubMed]
125. Dubowchik, G.M.; Walker, M.A. Receptor-mediated and enzyme-dependent targeting of cytotoxic anticancer drugs. *Pharmacol. Ther.* **1999**, *83*, 67–123. [CrossRef]
126. Kratz, F.; Müller, I.A.; Ryppa, C.; Warnecke, A. Prodrug strategies in anticancer chemotherapy. *ChemMedChem* **2008**, *3*, 20–53. [CrossRef] [PubMed]
127. Anami, Y.; Yamazaki, C.M.; Xiong, W.; Gui, X.; Zhang, N.; An, Z.; Tsuchikama, K. Glutamic acid–valine–citrulline linkers ensure stability and efficacy of antibody–drug conjugates in mice. *Nat. Commun.* **2018**, *9*, 2512. [CrossRef]
128. Wei, B.; Gunzner-Toste, J.; Yao, H.; Wang, T.; Wang, J.; Xu, Z.; Chen, J.; Wai, J.; Nonomiya, J.; Tsai, S.P.; et al. Discovery of peptidomimetic antibody–drug conjugate linkers with enhanced protease specificity. *J. Med. Chem.* **2018**, *61*, 989–1000. [CrossRef]
129. Lyon, R.P.; Bovee, T.D.; Doronina, S.O.; Burke, P.J.; Hunter, J.H.; Neff-LaFord, H.D.; Jonas, M.; Anderson, M.E.; Setter, J.R.; Senter, P.D. Reducing hydrophobicity of homogeneous antibody-drug conjugates improves pharmacokinetics and therapeutic index. *Nat. Biotechnol.* **2015**, *33*, 733–735. [CrossRef]
130. Kolodych, S.; Michel, C.; Delacroix, S.; Koniev, O.; Ehkirch, A.; Eberova, J.; Cianferani, S.; Renoux, B.; Krezel, W.; Poinot, P.; et al. Development and evaluation of beta-galactosidase-sensitive antibody-drug conjugates. *Eur. J. Med. Chem.* **2017**, *142*, 376–382. [CrossRef]

131. DiJoseph, J.F.; Dougher, M.M.; Kalyandrug, L.B.; Armellino, D.C.; Boghaert, E.R.; Hamann, P.R.; Moran, J.K.; Damle, N.K. Antitumor efficacy of a combination of CMC-544 (Inotuzumab Ozogamicin), a CD22-targeted cytotoxic immunoconjugate of calicheamicin, and rituximab against Non-Hodgkin's B-Cell lymphoma. *Clin. Cancer Res.* **2006**, *12*, 242–249. [CrossRef]
132. Zhou, D.; Casavant, J.; Graziani, E.I.; He, H.; Janso, J.; Loganzo, F.; Musto, S.; Tumey, N.; O'Donnell, C.J.; Dushin, R. Novel PIKK inhibitor antibody-drug conjugates: Synthesis and anti-tumor activity. *Bioorg. Med. Chem. Lett.* **2019**, *29*, 943–947. [CrossRef]
133. Govindan, S.V.; Cardillo, T.M.; Sharkey, R.M.; Tat, F.; Gold, D.V.; Goldenberg, D.M. Milatuzumab-SN-38 conjugates for the treatment of CD74+ cancers. *Mol. Cancer Ther.* **2013**, *12*, 968–978. [CrossRef]
134. Kellogg, B.A.; Garrett, L.; Kovtun, Y.; Lai, K.C.; Leece, B.; Miller, M.; Payne, G.; Steeves, R.; Whiteman, K.R.; Widdison, W.; et al. Disulfide-linked antibody−maytansinoid conjugates: Optimization of in vivo activity by varying the steric hindrance at carbon atoms adjacent to the disulfide linkage. *Bioconjug. Chem.* **2011**, *22*, 717–727. [CrossRef] [PubMed]
135. Bernardes, G.J.L.; Casi, G.; Trüssel, S.; Hartmann, I.; Schwager, K.; Scheuermann, J.; Neri, D. A traceless vascular-targeting antibody–drug conjugate for cancer therapy. *Angew. Chem. Int. Ed.* **2012**, *51*, 941–944. [CrossRef] [PubMed]
136. Baldwin, A.D.; Kiick, K.L. Tunable degradation of maleimide—Thiol adducts in reducing environments. *Bioconjug. Chem.* **2011**, *22*, 1946–1953. [CrossRef] [PubMed]
137. Yasunaga, M.; Manabe, S.; Matsumura, Y. Immunoregulation by IL-7R-targeting antibody-drug conjugates: Overcoming steroid-resistance in cancer and autoimmune disease. *Sci. Rep.* **2017**, *7*, 10735. [CrossRef]
138. Hampe, C.S. Protective role of anti-idiotypic antibodies in autoimmunity—Lessons for type 1 diabetes. *Autoimmunity* **2012**, *45*, 320–331. [CrossRef] [PubMed]
139. Senter, P.D. Potent antibody drug conjugates for cancer therapy. *Curr. Opin. Chem. Biol.* **2009**, *13*, 235–244. [CrossRef]
140. Lewis Phillips, G.D.; Li, G.; Dugger, D.L.; Crocker, L.M.; Parsons, K.L.; Mai, E.; Blättler, W.A.; Lambert, J.M.; Chari, R.V.J.; Lutz, R.J.; et al. Targeting HER2-positive breast cancer with trastuzumab-DM1, an antibody–cytotoxic drug conjugate. *Cancer Res.* **2008**, *68*, 9280–9290. [CrossRef]
141. Kraynov, E.; Kamath, A.V.; Walles, M.; Tarcsa, E.; Deslandes, A.; Iyer, R.A.; Datta-Mannan, A.; Sriraman, P.; Bairlein, M.; Yang, J.J.; et al. Current approaches for absorption, distribution, metabolism, and excretion characterization of antibody-drug conjugates: An industry white paper. *Drug Metab. Dispos.* **2016**, *44*, 617–623. [CrossRef]
142. Drake, P.M.; Rabuka, D. Recent developments in ADC technology: Preclinical studies signal future clinical trends. *BioDrugs* **2017**, *31*, 521–531. [CrossRef]
143. Donaghy, H. Effects of antibody, drug and linker on the preclinical and clinical toxicities of antibody-drug conjugates. *mAbs* **2016**, *8*, 659–671. [CrossRef]
144. Panowski, S.; Bhakta, S.; Raab, H.; Polakis, P.; Junutula, J.R. Site-specific antibody drug conjugates for cancer therapy. *mAbs* **2014**, *6*, 34–45. [CrossRef] [PubMed]
145. Shah, D.K.; Betts, A.M. Antibody biodistribution coefficients. *mAbs* **2013**, *5*, 297–305. [CrossRef] [PubMed]
146. Yip, V.; Palma, E.; Tesar, D.B.; Mundo, E.E.; Bumbaca, D.; Torres, E.K.; Reyes, N.A.; Shen, B.Q.; Fielder, P.J.; Prabhu, S.; et al. Quantitative cumulative biodistribution of antibodies in mice. *mAbs* **2014**, *6*, 689–696. [CrossRef] [PubMed]
147. Herbertson, R.A.; Tebbutt, N.C.; Lee, F.T.; MacFarlane, D.J.; Chappell, B.; Micallef, N.; Lee, S.T.; Saunder, T.; Hopkins, W.; Smyth, F.E.; et al. Phase I biodistribution and pharmacokinetic study of Lewis Y-targeting immunoconjugate CMD-193 in patients with advanced epithelial cancers. *Clin. Cancer Res.* **2009**, *15*, 6709–6715. [CrossRef]
148. Lu, D.; Joshi, A.; Wang, B.; Olsen, S.; Yi, J.H.; Krop, I.E.; Burris, H.A.; Girish, S. An integrated multiple-analyte pharmacokinetic model to characterize trastuzumab emtansine (T-DM1) clearance pathways and to evaluate reduced pharmacokinetic sampling in patients with HER2-positive metastatic breast cancer. *Clin. Pharm.* **2013**, *52*, 657–672. [CrossRef]
149. Han, T.H.; Gopal, A.K.; Ramchandren, R.; Goy, A.; Chen, R.; Matous, J.V.; Cooper, M.; Grove, L.E.; Alley, S.C.; Lynch, C.M.; et al. CYP3A-mediated drug-drug interaction potential and excretion of brentuximab vedotin, an antibody-drug conjugate, in patients with CD30-positive hematologic malignancies. *J. Clin. Pharmacol.* **2013**, *53*, 866–877. [CrossRef]

150. Collins, D.M.; Bossenmaier, B.; Kollmorgen, G.; Niederfellner, G. Acquired resistance to antibody-drug conjugates. *Cancers* **2019**, *11*, 394. [CrossRef]
151. Bobaly, B.; Fleury-Souverain, S.; Beck, A.; Veuthey, J.L.; Guillarme, D.; Fekete, S. Current possibilities of liquid chromatography for the characterization of antibody-drug conjugates. *J. Pharm. Biomed. Anal.* **2018**, *147*, 493–505. [CrossRef]
152. Lechner, A.; Giorgetti, J.; Gahoual, R.; Beck, A.; Leize-Wagner, E.; Francois, Y.N. Insights from capillary electrophoresis approaches for characterization of monoclonal antibodies and antibody drug conjugates in the period 2016–2018. *J. Chromatogr. B* **2019**, *1122*, 1–17. [CrossRef]
153. Wagh, A.; Song, H.; Zeng, M.; Tao, L.; Das, T.K. Challenges and new frontiers in analytical characterization of antibody-drug conjugates. *mAbs* **2018**, *10*, 222–243. [CrossRef]
154. Beck, A.; D'Atri, V.; Ehkirch, A.; Fekete, S.; Hernandez-Alba, O.; Gahoual, R.; Leize-Wagner, E.; Francois, Y.; Guillarme, D.; Cianferani, S. Cutting-edge multi-level analytical and structural characterization of antibody-drug conjugates: Present and future. *Expert Rev. Proteom.* **2019**, *16*, 337–362. [CrossRef] [PubMed]
155. Yurkovetskiy, A.V.; Yin, M.; Bodyak, N.; Stevenson, C.A.; Thomas, J.D.; Hammond, C.E.; Qin, L.; Zhu, B.; Gumerov, D.R.; Ter-Ovanesyan, E.; et al. A polymer-based antibody-vinca drug conjugate platform: Characterization and preclinical efficacy. *Cancer Res.* **2015**, *75*, 3365–3372. [CrossRef] [PubMed]
156. Goldenberg, D.M.; Cardillo, T.M.; Govindan, S.V.; Rossi, E.A.; Sharkey, R.M. Trop-2 is a novel target for solid cancer therapy with sacituzumab govitecan (IMMU-132), an antibody-drug conjugate (ADC). *Oncotarget* **2015**, *6*, 22496–22512. [CrossRef] [PubMed]
157. Viricel, W.; Fournet, G.; Beaumel, S.; Perrial, E.; Papot, S.; Dumontet, C.; Joseph, B. Monodisperse polysarcosine-based highly-loaded antibody-drug conjugates. *Chem. Sci.* **2019**, *10*, 4048–4053. [CrossRef] [PubMed]
158. Schneider, H.; Deweid, L.; Pirzer, T.; Yanakieva, D.; Englert, S.; Becker, B.; Avrutina, O.; Kolmar, H. Dextramabs: A novel format of antibody-drug conjugates featuring a multivalent polysaccharide scaffold. *ChemistryOpen* **2019**, *8*, 354–357. [CrossRef]
159. Wang, L.; Amphlett, G.; Blattler, W.A.; Lambert, J.M.; Zhang, W. Structural characterization of the maytansinoid-monoclonal antibody immunoconjugate, huN901-DM1, by mass spectrometry. *Protein Sci.* **2005**, *14*, 2436–2446. [CrossRef]
160. Cao, M.; De Mel, N.; Jiao, Y.; Howard, J.; Parthemore, C.; Korman, S.; Thompson, C.; Wendeler, M.; Liu, D. Site-specific antibody-drug conjugate heterogeneity characterization and heterogeneity root cause analysis. *mAbs* **2019**, *11*, 1–13. [CrossRef]
161. Mehta, G.; Scheinman, R.I.; Holers, V.M.; Banda, N.K. A new approach for the treatment of arthritis in mice with a novel conjugate of an Anti-C5aR1 antibody and C5 small interfering RNA. *J. Immunol.* **2015**, *194*, 5446–5454. [CrossRef]
162. Ross, P.L.; Wolfe, J.L. Physical and chemical stability of antibody drug conjugates: Current status. *J. Pharm. Sci.* **2016**, *105*, 391–397. [CrossRef]
163. Duerr, C.; Friess, W. Antibody-drug conjugates-stability and formulation. *Eur. J. Pharm. Biopharm.* **2019**, *139*, 168–176. [CrossRef]
164. Wakankar, A.A.; Feeney, M.B.; Rivera, J.; Chen, Y.; Kim, M.; Sharma, V.K.; Wang, Y.J. Physicochemical stability of the antibody-drug conjugate Trastuzumab-DM1: Changes due to modification and conjugation processes. *Bioconjug. Chem.* **2010**, *21*, 1588–1595. [CrossRef] [PubMed]
165. Gandhi, A.V.; Randolph, T.W.; Carpenter, J.F. Conjugation of emtansine onto trastuzumab promotes aggregation of the antibody-drug conjugate by reducing repulsive electrostatic interactions and increasing hydrophobic interactions. *J. Pharm. Sci.* **2019**, *108*, 1973–1983. [CrossRef] [PubMed]
166. Beckley, N.S.; Lazzareschi, K.P.; Chih, H.W.; Sharma, V.K.; Flores, H.L. Investigation into temperature-induced aggregation of an antibody drug conjugate. *Bioconjug. Chem.* **2013**, *24*, 1674–1683. [CrossRef]
167. Adem, Y.T.; Schwarz, K.A.; Duenas, E.; Patapoff, T.W.; Galush, W.J.; Esue, O. Auristatin antibody drug conjugate physical instability and the role of drug payload. *Bioconjug. Chem.* **2014**, *25*, 656–664. [CrossRef] [PubMed]
168. Guo, J.; Kumar, S.; Prashad, A.; Starkey, J.; Singh, S.K. Assessment of physical stability of an antibody drug conjugate by higher order structure analysis: Impact of thiol- maleimide chemistry. *Pharm. Res.* **2014**, *31*, 1710–1723. [CrossRef] [PubMed]

169. Guo, J.; Kumar, S.; Chipley, M.; Marcq, O.; Gupta, D.; Jin, Z.; Tomar, D.S.; Swabowski, C.; Smith, J.; Starkey, J.A.; et al. Characterization and higher-order structure assessment of an interchain cysteine-based ADC: Impact of drug loading and distribution on the mechanism of aggregation. *Bioconjug. Chem.* **2016**, *27*, 604–615. [CrossRef]
170. Buecheler, J.W.; Winzer, M.; Tonillo, J.; Weber, C.; Gieseler, H. Impact of payload hydrophobicity on the stability of antibody-drug conjugates. *Mol. Pharm.* **2018**, *15*, 2656–2664. [CrossRef]
171. Gandhi, A.V.; Arlotta, K.J.; Chen, H.N.; Owen, S.C.; Carpenter, J.F. Biophysical properties and heating-induced aggregation of lysine-conjugated antibody-drug conjugates. *J. Pharm. Sci.* **2018**, *107*, 1858–1869. [CrossRef]
172. Ohri, R.; Bhakta, S.; Fourie-O'Donohue, A.; Dela Cruz-Chuh, J.; Tsai, S.P.; Cook, R.; Wei, B.; Ng, C.; Wong, A.W.; Bos, A.B.; et al. High-throughput cysteine scanning to identify stable antibody conjugation sites for maleimide- and disulfide-based linkers. *Bioconjug. Chem.* **2018**, *29*, 473–485. [CrossRef]
173. Fathi, A.T.; Erba, H.P.; Lancet, J.E.; Stein, E.M.; Ravandi, F.; Faderl, S.; Walter, R.B.; Advani, A.S.; DeAngelo, D.J.; Kovacsovics, T.J.; et al. A phase 1 trial of vadastuximab talirine combined with hypomethylating agents in patients with CD33-positive AML. *Blood* **2018**, *132*, 1125–1133. [CrossRef]
174. King, G.T.; Eaton, K.D.; Beagle, B.R.; Zopf, C.J.; Wong, G.Y.; Krupka, H.I.; Hua, S.Y.; Messersmith, W.A.; El-Khoueiry, A.B. A phase 1, dose-escalation study of PF-06664178, an anti-Trop-2/Aur0101 antibody-drug conjugate in patients with advanced or metastatic solid tumors. *Investig. New Drugs* **2018**, *36*, 836–847. [CrossRef] [PubMed]
175. Phillips, T.; Barr, P.M.; Park, S.I.; Kolibaba, K.; Caimi, P.F.; Chhabra, S.; Kingsley, E.C.; Boyd, T.; Chen, R.; Carret, A.-S.; et al. A phase 1 trial of SGN-CD70A in patients with CD70-positive diffuse large B cell lymphoma and mantle cell lymphoma. *Investig. New Drugs* **2019**, *37*, 297–306. [CrossRef] [PubMed]
176. Saber, H.; Simpson, N.; Ricks, T.K.; Leighton, J.K. An FDA oncology analysis of toxicities associated with PBD-containing antibody-drug conjugates. *Regul. Toxicol. Pharmacol.* **2019**, *107*, 104429. [CrossRef] [PubMed]
177. Drake, P.M.; Carlson, A.; McFarland, J.M.; Banas, S.; Barfield, R.M.; Zmolek, W.; Kim, Y.C.; Huang, B.C.B.; Kudirka, R.; Rabuka, D. CAT-02-106, a site-specifically conjugated Anti-CD22 antibody bearing an MDR1-resistant maytansine payload yields excellent efficacy and safety in preclinical models. *Mol. Cancer Ther.* **2018**, *17*, 161–168. [CrossRef] [PubMed]
178. Buecheler, J.W.; Winzer, M.; Weber, C.; Gieseler, H. Oxidation-induced destabilization of model antibody-drug conjugates. *J. Pharm. Sci.* **2019**, *108*, 1236–1245. [CrossRef]
179. Cockrell, G.M.; Wolfe, M.S.; Wolfe, J.L.; Schoneich, C. Photoinduced aggregation of a model antibody-drug conjugate. *Mol. Pharm.* **2015**, *12*, 1784–1797. [CrossRef]
180. Liu, H.; Gaza-Bulseco, G.; Faldu, D.; Chumsae, C.; Sun, J. Heterogeneity of monoclonal antibodies. *J. Pharm. Sci.* **2008**, *97*, 2426–2447. [CrossRef]
181. Chen, T.; Su, D.; Gruenhagen, J.; Gu, C.; Li, Y.; Yehl, P.; Chetwyn, N.P.; Medley, C.D. Chemical de-conjugation for investigating the stability of small molecule drugs in antibody-drug conjugates. *J. Pharm. Biomed. Anal.* **2016**, *117*, 304–310. [CrossRef]
182. Wakankar, A.; Chen, Y.; Gokarn, Y.; Jacobson, F.S. Analytical methods for physicochemical characterization of antibody drug conjugates. *mAbs* **2011**, *3*, 161–172. [CrossRef]
183. Su, D.; Kozak, K.R.; Sadowsky, J.; Yu, S.-F.; Fourie-O'Donohue, A.; Nelson, C.; Vandlen, R.; Ohri, R.; Liu, L.; Ng, C.; et al. Modulating antibody–drug conjugate payload metabolism by conjugation site and linker modification. *Bioconjug. Chem.* **2018**, *29*, 1155–1167. [CrossRef]

Review

Discovery Strategies to Maximize the Clinical Potential of T-Cell Engaging Antibodies for the Treatment of Solid Tumors

Vladimir Voynov [1,*], Paul J. Adam [2], Andrew E. Nixon [1] and Justin M. Scheer [1]

1 Biotherapeutics Discovery, Boehringer Ingelheim Pharmaceuticals, Inc., 900 Ridgebury Road, Ridgefield, CT 06877, USA; andrew.nixon@boehringer-ingelheim.com (A.E.N.); justin.scheer@boehringer-ingelheim.com (J.M.S.)

2 Cancer Immunology & Immune Modulation, Boehringer Ingelheim RCV GmbH & Co KG, Dr. Boehringer-Gasse 5-11, 1121 Vienna, Austria; paul.adam@boehringer-ingelheim.com

* Correspondence: vladimir.voynov@boehringer-ingelheim.com

Received: 13 September 2020; Accepted: 11 November 2020; Published: 18 November 2020

Abstract: T-cell Engaging bispecific antibodies (TcEs) that can re-direct cytotoxic T-cells to kill cancer cells have been validated in clinical studies. To date, the clinical success with these agents has mainly been seen in hematologic tumor indications. However, an increasing number of TcEs are currently being developed to exploit the potent mode-of-action to treat solid tumor indications, which is more challenging in terms of tumor-cell accessibility and the complexity of the tumor microenvironment (TME). Of particular interest is the potential of TcEs as an immunotherapeutic approach for the treatment of non-immunogenic (often referred to as cold) tumors that do not respond to checkpoint inhibitors such as programmed cell death protein 1 (PD-1) and programmed death ligand 1 (PD-L1) antibodies. This has led to considerable discovery efforts for, firstly, the identification of tumor selective targeting approaches that can safely re-direct cytotoxic T-cells to cancer cells, and, secondly, bispecific antibodies and their derivatives with drug-like properties that promote a potent cytolytic synapse between T-cells and tumor cells, and in the most advanced TcEs, have IgG-like pharmacokinetics for dosing convenience. Based on encouraging pre-clinical data, a growing number of TcEs against a broad range of targets, and using an array of different molecular structures have entered clinical studies for solid tumor indications, and the first clinical data is beginning to emerge. This review outlines the different approaches that have been taken to date in addressing the challenges of exploiting the TcE mode-of-action for a broad range of solid indications, as well as opportunities for future discovery potential.

Keywords: T-cell engagers; bispecific antibodies; immunotherapy; oncology; antibody engineering; immunological synapse

1. Introduction

Within the last decade, therapeutic antibodies in the field of cancer immunotherapy have been used to establish a new paradigm for cancer treatment. This has mainly been driven by the clinical data and subsequent approval of several checkpoint inhibitors (CPI), and has led to more than two thousand ongoing clinical trials with these agents as monotherapy or in combination with other therapies [1]. The remarkable success of cytotoxic T-lympocyte-associated protein 4 (CTLA4), PD-1, and PD-L1 antibodies is due to their ability to antagonize immune cell checkpoint inhibitor proteins and 'release the brake' on the ability of a patient's immune system to fight off tumors [2–6]. However, despite the high initial promise of such agents, it is now clear that only a fraction of cancer patients are showing significant clinical benefit to such agents [7]. CPI-responsive patients typically have tumors that have a high mutational burden and can be recognized by the immune system as foreign, as evidenced by the presence of tumor infiltrating lymphocytes (TILs), specifically cluster of differentiation 3 (CD3)+, CD8+ and CD4+ T-cells. Non-immunogenic tumors make up the majority of tumors across cancer indications and have no or low numbers of TILs that recognize the tumor and cannot be boosted by CPIs. For these patients, other strategies must be employed to promote the patients' cytotoxic immune cells to recognize the tumor cells.

Two technologies have emerged that can re-direct cytotoxic T-cells, independent of their natural T-cell receptor (TCR) specificity, to tumor antigens: Chimeric Antigen Receptor T-cells (CAR-T) and T-cell Engaging bispecific antibodies (TcE). While both technologies aim to achieve a similar therapeutic effect, they are very different drug classes, with CAR-T being a cellular therapy, and TcEs protein drugs based on antibody fragments and/or soluble TCRs. Recent reviews have addressed the similarities and differences between CAR-T and T-cell Engagers [8,9]. The therapeutic approach with T-cell Engagers achieved clinical success with the approval and use of Blinatumomab for treatment of relapsed and refractory acute lymphoblastic leukemia [10]. This Bispecific T-cell Engager (BiTE) is composed of two scFv domains (one targeting CD19 on malignant B-cells and the other targeting CD3 on T-cells) connected by a linker, to induce a cytolytic synapse between a T-cell and a CD19-positive tumor cell [11]. Additional BiTEs are progressing in clinical development [12–15]; however, one drawback of BiTE molecules is their fast clearance with half-life of just a few hours, so they are administered by daily intravenous infusions.

Unlike hematologic tumors where the cancer cells often manifest themselves in the blood or tissues where lymphoid or myeloid cells are present, the majority of solid tumors have a more complex microenvironment that represents a greater challenge for cancer therapies [16–20]. In these cases, TcEs offer a unique opportunity by recruiting cytotoxic immune cells to the solid tumor, and once the tumor cells have been lysed there is a chain reaction involving T-cell activation, proliferation, and recruitment of other immune cells into the tumor microenvironment (TME). The presence of T-cells in the tumor environment may activate checkpoint mechanisms meaning that a combination of TcE and CPI could have synergistic therapeutic potential [18,21,22].

2. Clinical Use

Currently, there are no approved T-cell engagers for solid tumors. Catumaxomab (EpCAMxCD3), a prototypic version of a TcE based on a mouse–rat hybrid IgG was approved in the European Union in 2009 to treat EpCAM-positive malignant ascites [23]. However, this agent was subsequently withdrawn from the market in 2017 for commercial reasons, likely driven by the fact that the high immunogenicity of the non-human antibody backbone made it useful only in the single-dose acute ascites setting. Ertumaxomab (Her2xCD3) is another similar mouse–rat hybrid IgG bispecific molecule that was the subject of several clinical trials, but did not reach approval, again possibly due to immunogenicity, and the presence of the tumor target in normal tissues representing a toxicity risk.

Despite the fact that there is a growing number of clinical trials for TcE biotherapeutics targeting solid tumor indications, these are significantly fewer than with CAR-Ts, and far less than with CPIs (Figure 1a). Approximately a third of the clinical trials ongoing with TcEs are targeting solid tumor indications, which is similar to the ratio with CAR-Ts, while most of CPIs are towards solid tumors (Figure 1b).

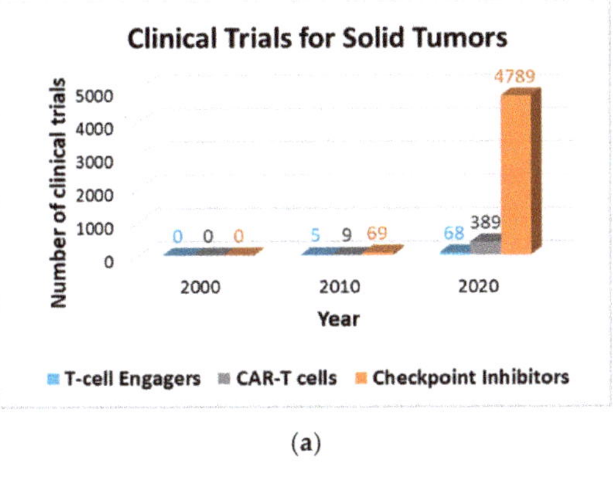

(**a**)

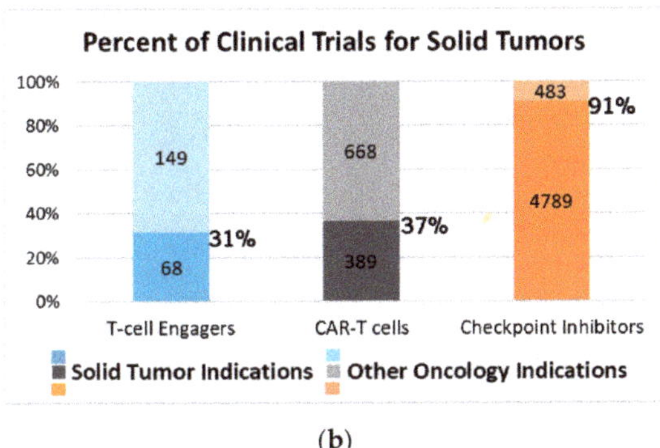

(**b**)

Figure 1. Clinical trials for solid tumors: (**a**) Comparison of number of clinical trials with TcEs, CAR-Ts and CPIs for solid tumors over time (2000, 2010, 2020). The numbers are from a search in the Citeline database by: (1) Mechanism of action: "CD3 agonism" for TcEs and "immune Checkpoint Inhibitors" for CPIs; (2) Therapeutic area: in both cases, "Oncology", and as sub-categories all listed solid tumor indications; (3) Therapeutic modality: monoclonal antibodies and all similar classes (chimeric, humanized, human); (4) Timeframe: before year 2000, between years 2000 and 2010, and in year 2020; (**b**) Percent of clinical trials for solid tumor indications vs. all oncology indications for each of the three therapeutic modalities. The numbers inside the columns correspond to the number of clinical trials. The percent number on the side of each column indicates the percent of clinical trials for solid tumor indications vs. for all oncology indications with each of these therapeutic modalities.

These lower numbers for TcEs towards solid tumors likely reflect the higher complexity and development time of the biotherapeutic modality. Two of the most advanced TcEs targeting solid tumors with ongoing clinical trials are IMCgp100 (Tebentafusp) and RO6958688 (Cibisatamab).

Tebentafusp showed partial responses and stable disease in several patients with uveal or cutaneous melanoma, and is currently in a Phase II clinical trial for metastatic uveal melanoma [24] (Table 1). This bispecific ImmTAC molecule comprises an affinity-optimized T-cell Receptor (TCR) domain that recognizes Human Leukocyte Antigen (HLA)-gp100peptide complex on tumor cells, and an anti-CD3 scFv that binds CD3 on T-cells to re-direct and activate the T-cells to lyse the gp100 positive tumor cells. One of the on-going clinical trials, NCT02535078, aims to evaluate the efficacy of Tebentafusp in combination with anti-PD-L1 and/or anti-CTLA-4 CPIs. Several other ImmTAC TcE molecules against targets presented in complex with HLA (NY-ESO-1, LAGE-1A, MAGE-A4, PRAME) are entering clinical trials (Table 1). These studies will not only test novel TcE technologies, but also broaden the therapeutic concept of targeting tumor-specific peptides in complex with HLA.

Table 1. Bispecific technologies and examples of specific T-cell Engager molecules in clinical trials for solid tumors. Features are derived from literature as described in the main text. Information on molecules and clinical trials is from clinicaltrials.gov and from Citeline. Of the 68 clinical trials in Figure 1, the molecules included in this table have clinical trial numbers assigned. The bispecific antibodies catumaxomab and ertumaxomab, for which all clinical trials are closed, are not included.

Technology and Key Features	Examples (Targets): Phase, Indication, Trial Number, Status (Other Information)
BiTE: -Two tandem scFvs; -short half-life (hours)	* **AMG 110/MT-110/solitomab (EpCAMxCD3):** -PhI, Solid tumors, NCT00635596, Completed * **MEDI-565/AMG 211/MT-111 (CEAxCD3):** -PhI, Gastrointestinal Adenocarcinomas, NCT01284231, Completed; -PhI, Advanced Gastrointestinal Cancer, NCT02291614, Completed (Terminated) * **Pasotuxizumab/AMG 212/MT-112/BAY 2010112 (PSMAxCD3):** -PhI, Prostate Cancer, NCT01723475, Completed (Terminated) * **AMG 596 (EGFRvIIIxCD3):** -PhI, Glioblastoma, NCT03296696, Recruiting (alone or in combination with AMG 404 (anti-PD-1))
ImmTAC: -Bispecific of a TCR domain and anti-CD3 scFv; -short half-life (hours)	* **Tebentafusp/IMCgp100 (gp100xCD3):** -Early PhI, Advanced Melanoma, NCT01209676, Completed; -PhI, Malignant Melanoma, NCT01211262, Completed; -PhII, Malignant Melanoma, NCT02889861, Terminated; -PhI/II, Malignant Melanoma, NCT02535078, Recruiting (combination with Durvalumab (anti-PD-L1) and/or Tremelimumab (anti-CTLA-4)); -PhI/II, Uveal Melanoma, NCT02570308, Active; -PhII, Uveal Melanoma, NCT03070392, Recruiting; * **IMCnieso (NY-ESO-1- and/or LAGE-1AxCD3):** -PhI/II, Advanced Solid Tumors, NCT03515551, Recruiting * **IMC-C103C (MAGE-A4xCD3):** -PhI/II, Advanced Solid Tumors, NCT03973333, Recruiting (alone and in combination with Atezolizumab (anti-PD-L1)) * **IMC-F106C (PRAMExCD3):** -PhI/II, Advanced Solid Tumors, NCT04262466, Recruiting (alone and in combination with CPIs)
TriTAC: -Trispecific construct: TAA-HSA-CD3, with anti-HSA binder for half-life extension	* **HPN424 (PSMAxCD3):** -PhI, Advanced Prostate Cancer, NCT03577028, Recruiting * **HPN536 (MesothelinxCD3):** -PhI/II, Advanced Cancers, NCT03872206, Recruiting
BiTE with Fc: -Two tandem scFvs linked to an Fc domain for half-life extension to several days, for less frequent dosing	* **AMG 160 (PSMAxCD3):** -PhI, Prostate Cancer, NCT03792841, Recruiting * **AMG 199 (MUC17xCD3):** -PhI, Gastric and Gastroesophageal Junction Cancers, NCT04117958, Recruiting * **AMG 757 (DLL3xCD3):** -PhI, Small Cell Lung Cancer, NCT03319940, Recruiting
Bispecific Antibody with common Light Chain: -Fab domain binders -common light chain -Fc domain for half-life extension -Fc mutations for heterodimerization of heavy chains	* **ERY974 (GPC3xCD3):** -PhI, Advanced Solid Tumors, NCT02748837, Completed; -PhI, JapicCTI-194805, Recruiting * **REGN4018 (MUC16xCD3):** -PhI/II, Recurrent Ovarian Cancer, NCT03564340, Recruiting (alone or in combination with Cemiplimab (anti-PD-1)
DuoBody Bispecific Antibody: -Fab domain binders -Fc domain for half-life extension -mutations in Fc -process for bispecific antibody generation from two regular IgGs after purification	* **JNJ-63898081 (PSMAxCD3):** -PhI, Advanced Stage Solid Tumors, NCT03926013, Recruiting

Table 1. *Cont.*

Technology and Key Features	Examples (Targets): Phase, Indication, Trial Number, Status (Other Information)
Bispecific TcE with Fc and bivalent for TAA: -Fc domain for half-life extension; -Knob-in-Hole technology in Fc for heterodimerization -CrossMab technology for correct LC-HC pairing in a bispecific -Two sites to bind TAA for improved therapeutic window.	* **Cibisatamab/RO6958688/RG7802 (CEAxCD3):** -PhI, Solid Tumors, NCT02324257, Completed; -PhI, Advanced Solid Tumors, JapicCTI-173764, Completed; -PhI, Solid Tumors, NCT02650713, Completed (in combination with Atezolizumab (anti-PD-L1)); -PhI/II, Non-small Cell Lung Cancer, NCT03337698, Recruiting; -PhI, Colorectal Cancer, NCT03866239, Active (in combination with Atezolizumab (anti-PD-L1) after pretreatment with Obinutuzumab (anti-CD20))
DART-Fc: -Fab or Fv domain binders with linkers -Fc domain for half-life extension -Monovalent or bivalent for targets	* **PF-06671008 (CDH3xCD3):** -PhI, Advanced Solid tumors, NCT02659631, Terminated * **MGD007 (gpA33xCD3):** -PhI, Colorectal Cancer, NCT02248805, Completed; -PhI/II, Metastatic Colorectal Cancer, NCT03531632, Active (in combination with MGA012 (anti-PD-1)) * **MGD009 (B7-H3xCD3):** -PhI, Solid Tumors, NCT02628535, Terminated; -PhI, Solid Tumors, NCT03406949, Recruiting (in combination with MGA012 (anti-PD-1)) * **PF07062119 (GUCY2CxCD3):** -PhI, Advanced or Metastatic Gastrointestinal Tumors, NCT04171141, Recruiting
Fab/scFv-Fc Bispecific monovalent (XmAb): -one binder is Fab; the other is scFv -Fc domain for half-life extension -engineered CH3 domain for heterodimerization	* **Tidutamab/XmAb18087 (SSTR2xCD3):** -PhI, Neuroendocrine and Gastrointestinal Stromal Tumors, NCT03411915, Recruiting * **GBR 1302/ISB 1302 (HER2xCD3):** -PhI, HER2+ Solid Tumors, NCT02829372, Terminated -PhI/II, Breast Cancer, NCT03983395, Recruiting * **AMG 509 (STEAP1xCD3):** -PhI, Prostate Cancer, NCT04221542, Recruiting * **M701 (EpCAMxCD3):** -PhI, Ascites, Solid Tumors, ChiCTR1900024144, Recruiting * **M802 (HER2xCD3):** -PhI, HER2+ Solid Tumors, ChiCTR1900024128, Recruiting
scFv-Fc-scFv bispecific bivalent: -scFv domain binders -Fc domain for half-life extension -Bispecific and bivalent for targets	* **ES414/APVO414/MOR209 (PSMAxCD3):** -PhI, Prostate Cancer, NCT02262910, Completed (Terminated)
Fab/scFv-Fc bispecific bivalent: -scFv for CD3 attached to the C-terminus of the light chain of IgG -Fc domain for half-life extension	* **Hu3F8-BsAb (GD2xCD3):** -PhI/II, Neuroblastoma, Osteosarcoma, Other Solid Tumors, NCT03860207, Recruiting;
Other	* **BTRC4017A/RG6194 (Her2xCD3):** -PhI, HER2+ Solid Tumors, NCT03448042, Recruiting * **GEM3PSCA (PSCAxCD3):** -PhI, Solid Tumors, NCT03927573, Recruiting * **REGN5678 (PSMAxCD28):** -PhI, Prostate Cancer, NCT03972657, Recruiting (in combination with Cemiplimab (anti-PD-1)) * **CCW702/ABBV-154 (PSMAxCD3):** -PhI, Prostate Cancer, NCT04077021, Recruiting * **AMV564 (CD33xCD3):** -PhI, Advanced Solid Tumors, NCT04128423, Recruiting * **A-337 (EpCAMxCD3):** -PhI, Advanced Solid Tumors, ACTRN12617001181392, Terminated

Cibisatamab is being evaluated in several clinical trials for treatment of patients with CEA-positive solid tumors, such as non-small cell lung cancer and colorectal cancer (Table 1). The molecule has three binding domains, two Fabs that contact the target and one Fab that binds to CD3. This 2 + 1 structure allows for avid binding to the tumor antigen for improved therapeutic window, while concomitantly engaging CD3 on T-cells [25]. The molecule features a Crossmab technology for correct light-to-heavy chain pairing in the Fabs, and knob-in-hole technology for heterodimerization of the Fc. One of the clinical trials, NCT03866239, is evaluating Cibisatamab in combination with anti-PD-L1 therapy, consistent with the premise of synergies between the TcE and CPI mechanisms of action. This clinical trial also includes pre-treatment of patients with obinutuzumab (anti-CD20) to prevent occurrence of anti-drug antibodies, observed in earlier studies.

Tens of other bispecific TcEs for solid tumors are in clinical trials (Table 1). Several molecules have not progressed, possibly because of insufficient efficacy, or of toxicities due to expression of the tumor associated antigen (TAA) also in healthy tissues, or to target-independent T-cell activation. There appears to be a trend towards half-life extended TcE modalities, and many of the clinical trials are of TcEs in combination with CPI therapy. Based on the so far limited clinical experience with TcE agents in solid tumors, how can we assure the clinical translation of the TcE programs currently in pre-clinical development? Do we have the necessary technologies, or do we need new technologies? Drug discovery strategies to maximize the potential clinical benefit of the TcE therapeutic approach for solid tumors are reviewed, discussed and proposed below.

3. Challenges

3.1. Targeting Strategies for Solid Tumors

A fundamental challenge for designing effective TcE therapies for solid tumors is the identification of tumor selective targeting antigens. The identification of such tumor associated antigens (TAAs) and lineages, and their utility for the targeting of therapeutic antibodies selectively to tumors, have been an area of intense research for many years [26,27]. Several of the TcE molecules in clinical trials use TAAs, such as Her2, that have been successfully targeted in the past with other therapeutic modalities such as regular IgGs or antibody-drug conjugates. A TcE modality offers a differentiated and arguably more efficacious approach. Compared to a regular IgG, T-cell redirected cytotoxicity is considered a more potent and efficacious approach to targeting solid tumors than Fc-mediated antibody-dependent cell-mediated cytotoxicity (ADCC). Compared with antibody drug conjugates (ADC), T-cell redirected cytotoxicity relies on the host's immune system rather than on conjugation to cytotoxic chemical payloads, and it attacks dormant as well as actively dividing cancer cells. However, targets like Her2, EpCAM, CEA, are challenging to target with the highly potent TcE mechanism of action because of basal expression on healthy tissue, despite overexpression in tumor cells.

In more recent years, detailed analysis of the transcriptome, proteome, and metabolome of diseased versus healthy cells and tissues has been used to identify tumor-selective targeting proteins [28]. One example of a cell-surface target that has emerged as highly tumor-specific is Delta-like Ligand 3 (DLL3) [29]. Earlier, an ADC approach was used, but a Phase II clinical trial with Rovalpituzumab tesirine in DLL3-expressing small-cell lung cancer did not show significant overall benefit [30]. More recently, a TcE approach was pursued towards DLL3-positive small-cell lung cancer [31,32], and the next few years will indicate whether the TcE modality confers good translation of very promising pre-clinical results into the clinic.

In another example, a recent peptidome study using mass spectrometry analysis identified HLA-complexed peptides and developed a predictive tool for neo-antigens [33]. This and other studies have led to innovative TcE approaches targeting tumor selective major histocompatibility complex (MHC)/peptide complexes [34]. The advantage of targeting tumor presented peptide antigens is that it opens up the tumor selective protein space to include intracellular proteins that could not normally be targeted with an antibody approach. Tebentafusp is the leading TcE example that uses this target class, and others have now started clinical development.

To date, the most successful approach to achieve tumor cell target-dependent activation of T-cells is via the targeting of unique epitopes on the T-cell receptor, CD3. The CD3 targeting arm of current TcEs is often derived from one of two binders identified in the 1970–80's that bind to the CD3ε subunit, OKT3 [35] and SP34 [36,37]. Since then, different sequence optimized versions of these mouse-derived antibodies have been generated, for example to reduce the risk of immunogenicity in patients. The current thinking on CD3 affinity is that binding to CD3 with nM range affinities is advantageous over pM binding, with the expectation that weaker binding to CD3 would be less likely to cause TAA-independent T-cell activation and lysis. More recently, a screening approach was used to identify new, improved CD3 binders [38]—when formatted into a bispecific, some of these binders against different CD3 epitopes confer strong tumor cell killing, with minimal cytokine release [38], which is considered important towards maximizing a therapeutic window in the clinic. Regarding valency to CD3, having only one anti-CD3 binding arm in the multispecific TcE molecule is thought to result in a better safety profile because of lower risk of off-tumor T-cell activation. However, among all the different TcE programs that are pursued currently, there are also examples of molecules that are bivalent for CD3. In the next few years, results on the different formats and binders may give a clearer picture on the optimal options of affinities, valency, and formats.

There is also a clinical example of a TcE that uses CD28 instead of CD3 as the activating target receptor on T-cells (PSMAxCD28, NCT03972657). In addition, a recent preclinical study of co-activating CD3 and CD28 with a trispecific molecule targeting CD38 has shown significantly more potency in in vitro assays than the corresponding bispecific constructs that activate either CD3 or CD28 [39]. However, a control trispecific molecule without a CD38 binding arm also shows very strong activity, indicating that target dependent activation has been compromised. This illustrates the fine balance between potency and specificity that TcEs need to strike for optimal therapeutic benefit.

Beyond the identification of suitable tumor targeting antigens to enable TcEs, pre-clinical pharmacological analysis includes the evaluation of TcEs in in vitro potency assays and in vivo efficacy models of the disease to identify the most promising therapeutic candidates. Elegant cell-based functional assays allow evaluation of multiple target binders and formats for T-cell activation and in vitro target cell lysis [12,21,40–44]. Patient-derived organoid assays also help bridge cell-based functional assays with pre-clinical in vivo efficacy studies and clinical trials, as described for CEA-positive solid tumors [45]. Similarly, the establishment of elaborate in vivo models facilitate the pre-clinical evaluation and ranking of candidate therapeutics. One successfully used disease-relevant model for TcEs being developed for the treatment of solid tumors uses so called 'immune avatar models' which consist of immune-deficient mice to establish a human tumor xenograft, followed by engraftment of human T-cells and then administration of the TcE molecule [12,29]. Other options for models include the use of transgenic mice that express human CD3ε, substituting the need to use immune-deficient mice and to engraft human T-cells [21], although one challenge for these models is that the human target has to be introduced into the mouse tumor cells if the target binder is not mouse cross-reactive. Ueda et al. [46] constructed a human CD3 transgenic mouse with all three CD3 subunits (CD3ε, CD3δ, CD3γ) replaced, and such in vivo model was applied to the evaluation of a Glypican3 (GPC3)/CD3 TcE, ERY974 [47]. Alternatively, bispecific molecules that bind mouse CD3 can be evaluated in fully immune-competent mouse models. The disadvantage of using the murine immune system for such models is that the human tumors cannot be engrafted, and a syngeneic tumor has to be used meaning that a surrogate binder has often to be created to the murine antigen target. In one such approach, Benonisson et al. identified that a CD3-bispecific TcE recruits several immune cell types to the tumor microenvironment in a syngeneic mouse model of melanoma [48].

It remains difficult to capture all the heterogeneity and complexity of the TME of patients into pre-clinical models of disease. However, better understanding of the complex biology of solid tumors with respect to any cellular or extracellular matrix barriers (for example, stromal cells or collagen, hyaluronic acid and fibronectin-rich matrix) also contributes to the design of better immuno-oncology model systems and therapies [49]. The increased knowledge of cell types, extracellular matrix components and their interplay in the TME permits the design of well-controlled experiments that ultimately can be more predictive of positive outcome in patients. In the next few years, we will learn how well the promising pre-clinical results on several TcE programs translate in the clinic.

3.2. Identifying Optimal Target Binders

Usually, the next step in the biologics discovery process following the identification and validation of suitable targets is the generation of antibody binders that meet several requirements. Sophisticated discovery platforms of synthetic libraries or humanized animals add to the more traditional immunization campaigns for antibody generation. Increased structural and computational capabilities also facilitate the identification of diverse set of binders, even to challenging molecular targets or epitopes. As with any other therapeutic target or modality, binding affinity is a very important criterion. In the case of TcEs, another key parameter is epitope. Species cross-reactivity and biophysical properties further define the evaluation of binders to identify the optimal ones for TcEs.

Affinity is one important consideration for potent TcEs. Strong binding ($K_D < 1$ nM) to the TAA is considered a pre-requisite, especially for targets with a very low copy number on the surface of the tumor cells, and a notable example is Tebentafusp. Even though identification of strong, selective, binders to specific MHC-complexed peptides is challenging due to the nature of the complexed peptide, Tebentafusp exhibits a pM binding affinity [24]. Meanwhile, for a TAA that also has low expression levels on healthy normal tissues, it is possible to achieve a therapeutic window using avidity optimized weak binders (K_D 1–100 nM), where the molecule binds preferentially to higher-target expressing tumor cells, and sparing the lower-target expressing normal cells. For example, a bispecific format with bivalent Her2 binding improved the selectivity towards Her2-positive tumor cells over healthy cells through avidity [44]. Cibisatamab (CEAxCD3) is another molecule in such a 2:1 format, with two binding arms for the TAA, and one for CD3.

Epitope on the target is another key parameter for conferring strong TcE potency. A study by Bluemel et al. demonstrated that cell-membrane proximity of the epitope determines the potency of BiTEs, especially for large surface antigens such as melanoma chondroitin sulfate proteoglycan (MCSP) [40]. A similar observation was reported for efficient synapse formation with TcEs directed to a membrane-proximal epitope on another TAA, FcRH5, a B-cell lineage marker [50]. There may be other requirements for epitope selection, for example in cases where regions are bound by natural ligands, or due to sequence homology to other family members, or across other species. Ultimately, the epitope space for a given TAA might be significantly restricted; however, novel antibody generation strategies, as well as new discovery platforms, enable identification of binders to unique, previously inaccessible epitopes.

Species cross-reactivity is often another important criterion in lead identification. For example, binding cross-reactivity to the corresponding antigen in non-human primate species, such as cynomolgus monkey, enables important PK and safety studies before final molecule selection and clinical trials. Meanwhile, cross-reactivity to mouse or other species with disease-relevant models is important to be able to evaluate lead candidates in in vivo efficacy studies. If such cross-reactivity is not achieved or feasible, important in vivo parameters should be evaluated pre-clinically using a surrogate molecule and extrapolated to the lead therapeutic molecules. Human and cynomolgus CD3ε shows high sequence homology only in the first 30 amino acids, so this presents very limited epitope space to achieve human/cynomolgus species cross-reactivity, and human/mouse CD3ε sequences are even more divergent. Currently there are no known human/mouse CD3ε cross-reactive binders, and the only anti-mouseCD3 binder is 2C11 [51,52].

CMC properties (chemistry, manufacturing and controls) of the binders present an underlying objective throughout the discovery process. Even in regular IgGs, the Fab regions can have a profound effect on overall molecule stability and manufacturability, due to melting temperature, hydrophobicity and other biophysical and chemical properties of the CDRs and frameworks in the variable regions. Evaluation of the CMC properties of different binders is even more important in multispecific antibodies because of multiple variable domains, usage of fragment structures and non-IgG elements such as linkers and point variants for heterodimerization. High stability is incorporated in the design of recent Fab and non-Fab (scFv, VHH domains) synthetic libraries for antibody generation, by using some of the most common and most stable human germlines [53,54]. In addition, antibody libraries in non-Fab modalities are very useful for subsequent bispecific formatting with scFv and VHH components, to avoid any loss of binding or stability attributes during Fab to scFv or Fab to VHH engineering. There is an expanding set of biophysical techniques [55–57] and in silico tools [58–60] for developability assessment and engineering designs, many of which can be used in a high-throughput manner. Early manufacturability and biophysical evaluation during lead identification would advance more stable binders for bispecific TcE designs.

3.3. Multispecific Engineering Approaches

One of the drawbacks to the pioneering BiTE technology was the need for continuous intravenous infusion. BiTEs proved that TcEs can be developed and commercialized. The challenge was to improve on the drug-like properties, and this has been done by addressing several aspects of the technology. Early engineering approaches to improve the biophysical stability of Fvs established the use of linkers or interchain disulfide bonds [61,62], and enabled BiTE structures. Meanwhile, Knob-in-hole mutations in the CH3 domain of the Fc were ingeniously designed to form Fc heterodimers [63], and introduced Fc-containing, more antibody-like bispecifics. Subsequent to BiTEs, further protein engineering has led to various multispecific structures such as CrossMabs with CH1-CL crossover [64], TandAbs (Tandem diabodies) [65], DARTs (dual-affinity re-targeting) [66], ITEs (IgG-like T-cell Engaging bispecific antibody) [31], BEATs (Bispecific Engagement by Antibodies based on the T cell receptor), ImmTACs (Immune mobilising monoclonal TCRs against cancer) [34], TriTACs (Tri-specific T-cell Activating Construct) [67], and numerous other small domain or full antibody-like constructs [68,69]. Several such multispecific structures are currently used in TcE modality (Figure 2, Table 1). This variety of formats permits the identification of the most potent, safe and manufacturable ones for a given therapeutic concept [25,65,70–73]: monovalent vs. bivalent binding for one or both targets, different affinity and epitope binders, different size, distance and geometry of the binding domains and the whole molecule.

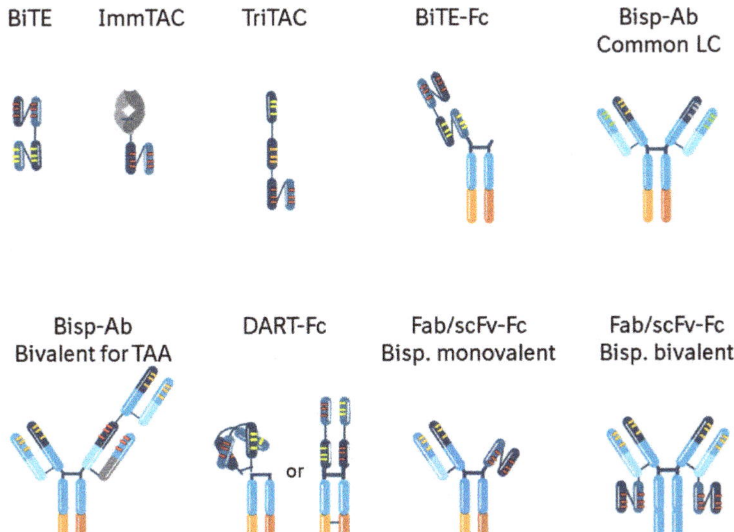

Figure 2. Examples of some of the Bispecific structures of T-cell Engager molecules in clinical trials. CDRs in the variable light and variable heavy chains for the same target are in the same color. Knob-in-Hole and other CH3 engineering technologies in the CH3 domain of the Fc are indicated by the different shade of orange.

Another example of important protein engineering for TcEs in light of the biology of solid tumors is the design of TME conditionally active molecules. Three TME conditions that have been explored so far are presence of specific metaloproteases, increased levels of ATP, or acidic pH. Probody is one of the first examples of a pro-drug antibody for improved therapeutic window [74]. A Probody has the antibody binding regions masked with a peptide which is processed by TME-specific proteases to then allow binding of the antibody to its target. At least four antibodies using Probody technology are in clinical trials as immunotherapy for solid tumors and also certain lymphomas [75]. The Probody concept was more recently applied to an EGFRxCD3 TcE, and preclinical studies indicate a more than 60-fold increase of maximum tolerated dose compared to unmasked bispecific construct [76]. Regarding ATP-dependent effects, Switch Antibody technology confers binding to a TAA, only in the presence of ATP, in the TME, as demonstrated for an anti-CD137 antibody [77]—the technology improves the safety profile for this target, and can be applied to other targets and platforms, including T-cell Redirecting Antibodies. Meanwhile, BioAlta describes low-pH specific binding to CD3 as part of their TcE and CAR-T platform for no off-tumor T-cell activation. All these examples of conditionally active TcEs are particularly relevant for TAAs that exhibit some basal expression in healthy tissues, and can be applied with respect to either the TAA or the T-cell antigen. Such technological advances can make more targets available and specifically help expand the usefulness of TcEs in solid tumors.

Safety considerations guide the design and characterization of TcE molecules in several additional aspects. The affinity and valency of the CD3 binder is important to assure that there is no target-independent T-cell activation, which in its worst manifestation could lead to cytokine release syndrome, and there are well-established in vitro T-cell activation assays for early screening of molecules. In addition, use of engineered Fc variants with weaker or no binding to Fcγ receptors results in significantly reduced effector function, and avoids potential undesirable cross-linking interactions of different immune cells [78,79]. Because of the presence of non-native elements such as linkers, non-Fab binding domains, swapped domains, and point variants in bispecific formats, the risk of immunogenicity of TcE molecules is higher than for regular IgGs. The multifaceted relationship of immunogenicity to sequence, stability of the molecule, antibody–target complex and higher-order structures, requires monitoring of immunogenicity and potential occurrence of anti-drug antibodies at all stages of pre-clinical and clinical investigation.

Regarding potency, several at least perceived challenges of TcEs for solid tumor indications include tumor penetration and efficiency in forming a strong immune synapse. Super-resolution and fluorescence microscopy allow a better understanding of the requirements for formation of a strong immune synapse for T-cell activation [80,81]. Meanwhile, a study with natural killer cells demonstrates that dextrans less than 4 nm are not limited in entering/exiting the immune synapse, while molecules around 10–13 nm are more than 50% impeded and dextrans greater than 32 nm are completely blocked [82]. Regular antibodies with about 150 kDa molecular weight have a hydrodynamic radius of 5–6 nm. It could be expected that smaller multispecific formats (BiTEs, ImmTACs and TriTACs are about 50 kDa) are more efficient than larger formats (about 150–200 kDa). However, based on pre-clinical in vivo studies on various TAAs, TcEs of both BiTE and bulkier HLE formats are able to achieve tumor growth inhibition and even regression with very low doses [21,25,31,41,42]. The size of the current bispecific TcE molecules does not seem to be an impediment to entering and forming a strong immune synapse.

Perhaps one of the biggest challenges for the development of bispecific antibody modalities has been the ability to ensure commercial manufacturing. In addition to identifying binders that have variable regions with good CMC properties, it is important that the complete bispecific TcE molecules are manufacturable and stable. There are a large number of multispecific formats currently available [68,71], and each of these has a unique set of manufacturability challenges, beyond platform processes for regular IgG molecules. Smaller formats such as BiTEs and ImmTACs do not allow Protein A-based affinity purification and do not benefit from molecule-stabilizing effects of antibody constant domains. On the other hand, larger biologics formats usually have an Fc domain that enables Protein A affinity purification, and confers additional stability and longer half-life. However, more often than not, TcE molecules with an Fc domain are asymmetric molecules, with different heavy and/or light chains. Expression and purification of such bispecific molecules is more challenging than that of regular IgG molecules, usually with lower expression level, lower initial purity after Protein A purification, and more polishing steps. So additional resources and time are required to build multispecific molecules with favorable CMC properties and to establish a robust manufacturing process for novel formats.

From the perspective of pharmacokinetics, a desired improvement in next-generation TcEs is the half-life of the therapeutic molecule, to allow better dosing convenience for patients. Because of their small size and domain composition (two scFvs connected by a linker), BiTE molecules clear very fast in vivo. With a short half-life of just a few hours, BiTE molecules have to be administered by continuous intravenous infusion. Several technologies, including BiTE-Fc fusions, IgG-like Fc containing formats, HSA/ABD fusion constructs, or PEGylation [83,84], enable longer half-life from at least several days to more than a week, and approaching the half-life of regular IgGs. As a result, in pre-clinical models of disease, high potency and efficacy can be achieved with once-weekly dosing over several weeks. Similarly, in the clinic, such HLE-TcEs are administered once weekly or less frequently as opposed to continuous infusion. While the PK profile of a biological drug is greatly improved by the presence of an Fc or an HSA-binding domain, the variable regions (especially if non-Fab format such as scFv or VHH) in a multispecific can significantly influence the in vivo stability and half-life of a molecule. In vitro serum stability and in vivo mouse PK studies can provide early information about which variable regions, linkers and formats are most suitable to advance for further testing.

4. Proposed Discovery Strategies and Conclusions

The complex and heterogeneous micro-environment of solid tumors means that for many cancer types, there remains a high unmet need for effective therapies. The clinical potential of the TcE mode-of-action has been demonstrated, and this is why it is important to continue to advance efforts in discovery and manufacturing to bring new generations of TcEs to patients, and address this high unmet medical need. Next generation TcEs should strike a balance of potency, safety, manufacturability and pharmacokinetics (Figure 3a). High potency depends a lot on the affinity of the binder, epitope, and the bispecific structure, in addition to target copy number on the target cells. Specificity and safety of the disease target as well as the therapeutic candidate are closely related, while multispecific engineering can offer additional target space even for targets with low expression on normal cells but significantly higher expression on tumor cells. Due to the more complex structure of bispecific molecules than regular antibodies, TcEs pose unique CMC challenges that have to be tackled to assure good manufacturability. Regarding pharmacokinetics, most of the current TcEs in clinical trials have a half-life extended profile allowing less frequent dosing, similar to that of regular IgGs.

Expertise from different disciplines addressing these key considerations provides a roadmap for a discovery strategy to benefit identification and evaluation of TcEs for solid tumors (Table 2), from biological understanding of the therapeutic challenges and opportunities of the TME, to the identification and validation of suitable TAAs, followed by the discovery of diverse set of binders, to be then incorporated in bispecific TcE molecules, for evaluation for function, safety, manufacturability and pharmacokinetics.

We propose a workflow of TcE discovery that includes parallel considerations of function, CMC properties, safety and PK profiling (Figure 3b). Improved technologies and capabilities permit the identification of diverse sets of affinity and epitope binders for a TAA, even for very challenging cell surface targets. Biophysical characterization of the binders alone and also in the context of bispecific structures identifies the molecules to advance for in vitro function assays, followed by mouse PK and in vivo efficacy studies. Mouse PK studies help determine the half-life and in vivo stability of the bispecific molecules, and can identify any challenges due to the targets, if the binders are cross-reactive, or due to variable region or engineered element sequences. Sequence optimization of the preferred binders and bispecific molecules is done to make the sequences as human as possible and to reduce critical quality attributes such as deamidation, aspartate isomerization, oxidation or fragmentation. The optimized lead candidates are subjected to a panel of rigorous pre-clinical testing for: (1) potency in in vitro and in vivo models of disease; (2) safety, half-life and in vivo stability in non-human primates; and (3) CMC properties. Favorable outcome of such comprehensive evaluation of TcE candidate therapeutics with good manufacturability and developability properties would assure a faster path to the clinic and more efficient approval and delivery to patients.

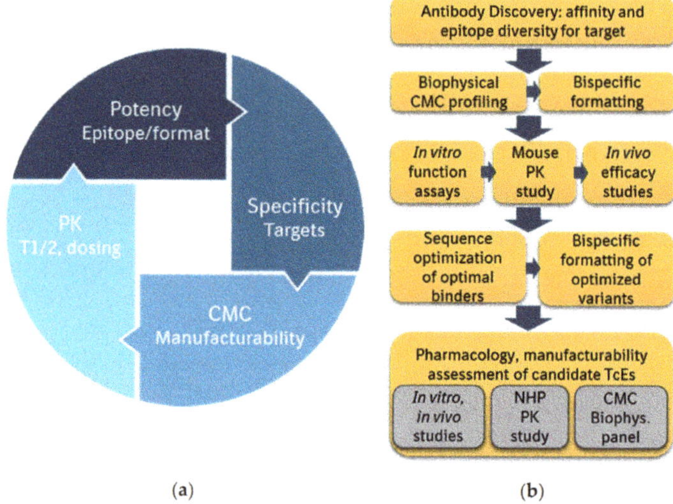

(**a**) (**b**)

Figure 3. T-cell Engagers in drug discovery: (**a**) Optimal features in a half-life extended Next-Gen T-cell Engager. First generation TcEs such as BiTEs show very high potency, but very short half-life, and challenging manufacturability. Current half-life extended TcEs aim to show strong potency, similar to BiTEs, while also featuring manufacturability and PK properties similar to regular IgG biologics; (**b**) Proposed workflow for drug discovery of T-cell Engagers. Discovery of antibodies with diverse affinity and epitopes to the target is beneficial, especially of novel targets. Side-by-side function and stability profiling are recommended before and after sequence-optimization, because of the interdependent importance of both Pharmacology and CMC for the clinical success of a TcE.

Table 2. Discovery strategy to benefit identification and evaluation of TcEs for solid tumors.

Subject	Key Considerations
TME Biology	Effectiveness of TcE modality for solid tumors Biomarkers and functional requirements of therapeutic molecule
Target Identification	Uniqueness of target for a therapeutic concept Expression profile in tumor vs. healthy cells and tissues
Lead Identification	Fab vs. non-Fab platforms for discovery of diverse set of binders Epitope, affinity, cross-reactivity, biophysical stability requirements
Multispecific formatting	Format that enables desired potency, safety, manufacturability and PK Evaluation of different binders in format for both function and CMC
CMC properties	Inherent molecule stability for optimal potency and safety Good manufacturability and developability for fast path to the clinic

Clinical and pre-clinical TcEs for solid tumors are against many different TAAs, and are built in various bispecific modalities. Each new entity brings a set of target-dependent and molecule-dependent challenges. Learnings from prior programs and advancements in biomedical research and development offer ways to address these challenges. The next few years promise to be of critical value in identifying optimal Immuno-oncology treatment options for solid tumor indications, where TcEs can make a unique and significant contribution, alone or in combination therapy.

Author Contributions: V.V., P.J.A., J.M.S. conceptualized and wrote the manuscript, and prepared the figures and tables, A.E.N. supported conceptualization and technologies towards binder discovery and characterization. All authors have read and agreed to the published version of the manuscript.

Funding: The authors are full time employees at Boehringer Ingelheim. This study was supported by the Basisprogramm grants of the Austrian Research Promotion Agency (FFG; 860968, 869530, and 875923).

Acknowledgments: The authors thank Patrizia Sini for help with the Citeline database search, David Young for help with Figure 2, and Sandeep Kumar, Susanne Hipp and Till Wenger for careful review and comments to the manuscript.

Conflicts of Interest: All authors declare the following conflict of interest with the contents of this article. All authors are employees of Boehringer Ingelheim affiliates. Boehringer Ingelheim discovers and develops T-cell engagers.

Abbreviations

ABD: Albumin binding domain; ADCC: antibody-dependent cell-mediated cytotoxicity; ADC: antibody-drug conjugate; ALL: acute lymphoblastic leukemia; BiTE: Bispecific T-cell Engager; CAR-T: chimeric antigen receptor T-cells; CMC: Chemistry, manufacturing and controls; CPI: checkpoint inhibitor; HLA: Human Leukocyte Antigen; HLE: Half-life extended (for molecule PK); HSA: Human serum albumin; MHC: major histocompatibility complex; PK: Pharmacokinetics; TAA: tumor associated antigen; TcE: T-cell Engaging bispecific antibodies; TcR: T-cell Receptor; TILs: Tumor infiltrating lymphocytes; TME: Tumor microenvironment.

References

1. Tang, J.; Yu, J.X.; Hubbard-Lucey, V.M.; Neftelinov, S.T.; Hodge, J.P.; Lin, Y. Trial watch: The clinical trial landscape for PD1/PDL1 immune checkpoint inhibitors. *Nat. Rev. Drug Discov.* **2018**, *17*, 854–855. [CrossRef]
2. Brown, J.A.; Dorfman, D.M.; Ma, F.-R.; Sullivan, E.L.; Munoz, O.; Wood, C.R.; Greenfield, E.A.; Freeman, G.J. Blockade of Programmed Death-1 Ligands on Dendritic Cells Enhances T Cell Activation and Cytokine Production. *J. Immunol.* **2003**, *170*, 1257–1266. [CrossRef] [PubMed]
3. Leach, D.R.; Krummel, M.F.; Allison, J.P. Enhancement of Antitumor Immunity by CTLA-4 Blockade. *Science* **1996**, *271*, 1734–1736. [CrossRef] [PubMed]
4. Nishimura, H.; Nose, M.; Hiai, H.; Minato, N.; Honjo, T. Development of Lupus-like Autoimmune Diseases by Disruption of the PD-1 Gene Encoding an ITIM Motif-Carrying Immunoreceptor. *Immunology* **1999**, *11*, 141–151. [CrossRef]
5. Hodi, F.S.; O'Day, S.J.; McDermott, D.F.; Weber, R.W.; Sosman, J.A.; Haanen, J.B.; Gonzalez, R.; Robert, C.; Schadendorf, D.; Hassel, J.C.; et al. Improved Survival with Ipilimumab in Patients with Metastatic Melanoma. *N. Engl. J. Med.* **2010**, *363*, 711–723. [CrossRef] [PubMed]
6. Topalian, S.L.; Hodi, F.S.; Brahmer, J.R.; Gettinger, S.N.; Smith, D.C.; McDermott, D.F.; Powderly, J.D.; Carvajal, R.D.; Sosman, J.A.; Atkins, M.B.; et al. Safety, Activity, and Immune Correlates of Anti–PD-1 Antibody in Cancer. *N. Engl. J. Med.* **2012**, *366*, 2443–2454. [CrossRef]
7. Haslam, A.; Prasad, V. Estimation of the Percentage of US Patients with Cancer Who Are Eligible for and Respond to Checkpoint Inhibitor Immunotherapy Drugs. *JAMA Netw. Open* **2019**, *2*, e192535. [CrossRef]
8. Slaney, C.Y.; Wang, P.; Darcy, P.K.P.; Kershaw, M.H. CARs versus BiTEs: A Comparison between T Cell–Redirection Strategies for Cancer Treatment. *Cancer Discov.* **2018**, *8*, 924–934. [CrossRef]
9. Strohl, W.R.; Naso, M. Bispecific T-Cell Redirection versus Chimeric Antigen Receptor (CAR)-T Cells as Approaches to Kill Cancer Cells. *Antibodies* **2019**, *8*, 41. [CrossRef]
10. Pulte, D.; Vallejo, J.; Przepiorka, D.; Nie, L.; Farrell, A.T.; Goldberg, K.B.; McKee, A.E.; Pazdur, R. FDA Supplemental Approval: Blinatumomab for Treatment of Relapsed and Refractory Precursor B-Cell Acute Lymphoblastic Leukemia. *Oncologist* **2018**, *23*, 1366–1371. [CrossRef]
11. Nagorsen, D.; Baeuerle, P.A. Immunomodulatory therapy of cancer with T cell-engaging BiTE antibody blinatumomab. *Exp. Cell Res.* **2011**, *317*, 1255–1260. [CrossRef] [PubMed]

12. Hipp, S.; Tai, Y.-T.; Blanset, D.; Deegen, P.; Wahl, J.; Thomas, O.; Rattel, B.; Adam, P.J.; Anderson, K.C.; Friedrich, M. A novel BCMA/CD3 bispecific T-cell engager for the treatment of multiple myeloma induces selective lysis in vitro and in vivo. *Leukemia* **2016**, *31*, 1743–1751. [CrossRef] [PubMed]
13. Topp, M.S.; Duell, J.; Zugmaier, G.; Attal, M.; Moreau, P.; Langer, C.; Krönke, J.; Facon, T.; Salnikov, A.V.; Lesley, R.; et al. Anti–B-Cell Maturation Antigen BiTE Molecule AMG 420 Induces Responses in Multiple Myeloma. *J. Clin. Oncol.* **2020**, *38*, 775–783. [CrossRef] [PubMed]
14. Rosenthal, M.; Balana, C.; Van Linde, M.E.; Sayehli, C.; Fiedler, W.M.; Wermke, M.; Massard, C.; Ang, A.; Kast, J.; Stienen, S.; et al. Novel anti-EGFRvIII bispecific T cell engager (BiTE) antibody construct in glioblastoma (GBM): Trial in progress of AMG 596 in patients with recurrent or newly diagnosed disease. *J. Clin. Oncol.* **2019**, *37*, TPS2071. [CrossRef]
15. Hummel, H.-D.; Kufer, P.; Grüllich, C.; Deschler-Baier, B.; Chatterjee, M.; Goebeler, M.-E.; Miller, K.; De Santis, M.; Loidl, W.C.; Buck, A.; et al. Phase I study of pasotuxizumab (AMG 212/BAY 2010112), a PSMA-targeting BiTE (Bispecific T-cell Engager) immune therapy for metastatic castration-resistant prostate cancer (mCRPC). *J. Clin. Oncol.* **2020**, *38*, 124. [CrossRef]
16. Giraldo, N.A.; Sanchez-Salas, R.; Peske, J.D.; Vano, Y.; Becht, E.; Petitprez, F.; Validire, P.; Ingels, A.; Cathelineau, X.; Fridman, W.H.; et al. The clinical role of the TME in solid cancer. *Br. J. Cancer* **2019**, *120*, 45–53. [CrossRef]
17. Angell, H.; Galon, J. From the immune contexture to the Immunoscore: The role of prognostic and predictive immune markers in cancer. *Curr. Opin. Immunol.* **2013**, *25*, 261–267. [CrossRef]
18. Galon, J.; Bruni, D. Approaches to treat immune hot, altered and cold tumours with combination immunotherapies. *Nat. Rev. Drug Discov.* **2019**, *18*, 197–218. [CrossRef]
19. Galon, J.; Costes, A.; Sanchez-Cabo, F.; Kirilovsky, A.; Mlecnik, B.; Lagorce-Pagès, C.; Tosolini, M.; Camus, M.; Berger, A.; Wind, P.; et al. Type, Density, and Location of Immune Cells Within Human Colorectal Tumors Predict Clinical Outcome. *Science* **2006**, *313*, 1960–1964. [CrossRef]
20. Whiteside, T.L. The tumor microenvironment and its role in promoting tumor growth. *Oncogene* **2008**, *27*, 5904–5912. [CrossRef]
21. Junttila, T.T.; Li, J.; Johnston, J.; Hristopoulos, M.; Clark, R.; Ellerman, D.; Wang, B.-E.; Li, Y.; Mathieu, M.; Li, G.; et al. Antitumor Efficacy of a Bispecific Antibody That Targets HER2 and Activates T Cells. *Cancer Res.* **2014**, *74*, 5561–5571. [CrossRef] [PubMed]
22. Kobold, S.; Pantelyushin, S.; Rataj, F.; Berg, J.V. Rationale for Combining Bispecific T Cell Activating Antibodies with Checkpoint Blockade for Cancer Therapy. *Front. Oncol.* **2018**, *8*, 285. [CrossRef] [PubMed]
23. Linke, R.; Klein, A.; Seimetz, D. Catumaxomab: Clinical development and future directions. *MAbs* **2010**, *2*, 129–136. [CrossRef] [PubMed]
24. Damato, B.; Dukes, J.; Goodall, H.; Carvajal, R.D. Tebentafusp: T Cell Redirection for the Treatment of Metastatic Uveal Melanoma. *Cancers* **2019**, *11*, 971. [CrossRef] [PubMed]
25. Bacac, M.; Fauti, T.; Sam, J.; Colombetti, S.; Weinzierl, T.; Ouaret, D.; Bodmer, W.F.; Lehmann, S.; Hofer, T.; Hosse, R.J.; et al. A Novel Carcinoembryonic Antigen T-Cell Bispecific Antibody (CEA TCB) for the Treatment of Solid Tumors. *Clin. Cancer Res.* **2016**, *22*, 3286–3297. [CrossRef] [PubMed]
26. Old, L.J. Cancer immunology: The search for specificity–G. H. A. Clowes Memorial lecture. *Cancer Res.* **1981**, *41*, 361–375.
27. Schietinger, A.; Philip, M.; Schreiber, K. Specificity in cancer immunotherapy. *Semin. Immunol.* **2008**, *20*, 276–285. [CrossRef]
28. Sengupta, S.; Sun, S.Q.; Huang, K.-L.; Oh, C.; Bailey, M.H.; Varghese, R.; Wyczalkowski, M.A.; Ning, J.; Tripathi, P.; McMichael, J.F.; et al. Integrative omics analyses broaden treatment targets in human cancer. *Genome Med.* **2018**, *10*, 1–20. [CrossRef]
29. Owen, D.H.; Giffin, M.J.; Bailis, J.M.; Smit, M.-A.D.; Carbone, D.P.; He, K. DLL3: An emerging target in small cell lung cancer. *J. Hematol. Oncol.* **2019**, *12*, 1–8. [CrossRef]
30. Morgensztern, D.; Besse, B.; Greillier, L.; Santana-Davila, R.; Ready, N.; Hann, C.L.; Glisson, B.S.; Farago, A.F.; Dowlati, A.; Rudin, C.M.; et al. Efficacy and Safety of Rovalpituzumab Tesirine in Third-Line and Beyond Patients with DLL3-Expressing, Relapsed/Refractory Small-Cell Lung Cancer: Results From the Phase II TRINITY Study. *Clin. Cancer Res.* **2019**, *25*, 6958–6966. [CrossRef]

31. Hipp, S.; Voynov, V.; Drobits-Handl, B.; Giragossian, C.; Trapani, F.; Nixon, A.E.; Scheer, J.M.; Adam, P.J. A Bispecific DLL3/CD3 IgG-like T-cell Antibody induces anti-tumor responses in Small Cell Lung Cancer. *Clin. Cancer Res.* **2020**, *26*, 5258–5268. [CrossRef] [PubMed]
32. Smit, M.-A.D.; Borghaei, H.; Owonikoko, T.K.; Hummel, H.-D.; Johnson, M.L.; Champiat, S.; Salgia, R.; Udagawa, H.; Boyer, M.J.; Govindan, R. Phase 1 study of AMG 757, a half-life extended bispecific T cell engager (BiTE) antibody construct targeting DLL3, in patients with small cell lung cancer (SCLC). *J. Clin. Oncol.* **2019**, *37*, TPS8577. [CrossRef]
33. Sarkizova, S.; Klaeger, S.; Le, P.M.; Li, L.W.; Oliveira, G.; Keshishian, H.; Hartigan, C.R.; Zhang, W.; Braun, D.A.; Ligon, K.L.; et al. A large peptidome dataset improves HLA class I epitope prediction across most of the human population. *Nat. Biotechnol.* **2019**, *38*, 199–209. [CrossRef] [PubMed]
34. Oates, J.; Jakobsen, K.B. ImmTACs: Novel bi-specific agents for targeted cancer therapy. *Oncoimmunology* **2013**, *2*, e22891. [CrossRef]
35. Kung, P.; Goldstein, G.; Reinherz, E.L.; Schlossman, S.F. Monoclonal antibodies defining distinctive human T cell surface antigens. *Science* **1979**, *206*, 347–349. [CrossRef]
36. Pessano, S.; Oettgen, H.; Bhan, A.K.; Terhorst, C. The T3/T cell receptor complex: Antigenic distinction between the two 20-kd T3 (T3-delta and T3-epsilon) subunits. *EMBO J.* **1985**, *4*, 337–344. [CrossRef]
37. Salmerón, A.; Sánchez-Madrid, F.; Ursa, M.A.; Fresno, M.; Alarcón, B. A conformational epitope expressed upon association of CD3-epsilon with either CD3-delta or CD3-gamma is the main target for recognition by anti-CD3 monoclonal antibodies. *J. Immunol.* **1991**, *147*, 3047–3052.
38. Trinklein, N.D.; Pham, D.; Schellenberger, U.; Buelow, B.; Boudreau, A.; Choudhry, P.; Clarke, S.C.; Dang, K.; Harris, K.E.; Iyer, S.; et al. Efficient tumor killing and minimal cytokine release with novel T-cell agonist bispecific antibodies. *mAbs* **2019**, *11*, 639–652. [CrossRef]
39. Wu, L.; Seung, E.; Xu, L.; Rao, E.; Lord, D.M.; Wei, R.R.; Cortez-Retamozo, V.; Ospina, B.; Posternak, V.; Ulinski, G.; et al. Trispecific antibodies enhance the therapeutic efficacy of tumor-directed T cells through T cell receptor co-stimulation. *Nat. Rev. Cancer* **2019**, *1*, 86–98. [CrossRef]
40. Bluemel, C.; Hausmann, S.; Fluhr, P.; Sriskandarajah, M.; Stallcup, W.B.; A Baeuerle, P.; Kufer, P. Epitope distance to the target cell membrane and antigen size determine the potency of T cell-mediated lysis by BiTE antibodies specific for a large melanoma surface antigen. *Cancer Immunol. Immunother.* **2010**, *59*, 1197–1209. [CrossRef]
41. Brischwein, K.; Schlereth, B.; Guller, B.; Steiger, C.; Wolf, A.; Lutterbuese, R.; Offner, S.; Locher, M.; Urbig, T.; Raum, T.; et al. MT110: A novel bispecific single-chain antibody construct with high efficacy in eradicating established tumors. *Mol. Immunol.* **2006**, *43*, 1129–1143. [CrossRef] [PubMed]
42. Fisher, T.S.; Hooper, A.T.; Lucas, J.; Clark, T.H.; Rohner, A.K.; Peano, B.; Elliott, M.W.; Tsaparikos, K.; Wang, H.; Golas, J.; et al. A CD3-bispecific molecule targeting P-cadherin demonstrates T cell-mediated regression of established solid tumors in mice. *Cancer Immunol. Immunother.* **2017**, *67*, 247–259. [CrossRef] [PubMed]
43. Löffler, A.; Gruen, M.; Wuchter, C.; Schriever, F.; Kufer, P.; Dreier, T.; Hanakam, F.; Baeuerle, P.A.; Bommert, K.; Karawajew, L.; et al. Efficient elimination of chronic lymphocytic leukaemia B cells by autologous T cells with a bispecific anti-CD19/anti-CD3 single-chain antibody construct. *Leukemia* **2003**, *17*, 900–909. [CrossRef] [PubMed]
44. Slaga, D.; Ellerman, D.; Lombana, T.N.; Vij, R.; Li, J.; Hristopoulos, M.; Clark, R.; Johnston, J.; Shelton, A.; Mai, E.; et al. Avidity-based binding to HER2 results in selective killing of HER2-overexpressing cells by anti-HER2/CD3. *Sci. Transl. Med.* **2018**, *10*, 5775. [CrossRef] [PubMed]
45. Gonzalez-Exposito, R.; Semiannikova, M.; Griffiths, B.; Khan, K.; Barber, L.J.; Woolston, A.; Spain, G.; Von Loga, K.; Challoner, B.; Patel, R.; et al. CEA expression heterogeneity and plasticity confer resistance to the CEA-targeting bispecific immunotherapy antibody cibisatamab (CEA-TCB) in patient-derived colorectal cancer organoids. *J. Immunother. Cancer* **2019**, *7*, 101. [CrossRef] [PubMed]
46. Ueda, O.; Wada, N.A.; Kinoshita, Y.; Hino, H.; Kakefuda, M.; Ito, T.; Fujii, E.; Noguchi, M.; Sato, K.; Morita, M.; et al. Entire CD3epsilon, delta, and gamma humanized mouse to evaluate human CD3-mediated therapeutics. *Sci. Rep.* **2017**, *7*, 45839. [CrossRef]
47. Ishiguro, T.; Sano, Y.; Komatsu, S.I.; Kamata-Sakurai, M.; Kaneko, A.; Kinoshita, Y.; Shiraiwa, H.; Azuma, Y.; Tsunenari, T.; Kayukawa, Y.; et al. An anti-glypican 3/CD3 bispecific T cell-redirecting antibody for treatment of solid tumors. *Sci. Transl. Med.* **2017**, *9*, 4291. [CrossRef]

48. Benonisson, H.; Altıntaş, I.; Sluijter, M.; Verploegen, S.; Labrijn, A.F.; Schuurhuis, D.H.; Houtkamp, M.A.; Verbeek, J.S.; Schuurman, J.; Van Hall, T. CD3-Bispecific Antibody Therapy Turns Solid Tumors into Inflammatory Sites but Does Not Install Protective Memory. *Mol. Cancer Ther.* **2018**, *18*, 312–322. [CrossRef]
49. Puré, E.; Lo, A. Can Targeting Stroma Pave the Way to Enhanced Antitumor Immunity and Immunotherapy of Solid Tumors? *Cancer Immunol. Res.* **2016**, *4*, 269–278. [CrossRef]
50. Li, J.; Stagg, N.J.; Johnston, J.; Harris, M.J.; Menzies, S.A.; DiCara, D.; Clark, V.; Hristopoulos, M.; Cook, R.; Slaga, D.; et al. Membrane-Proximal Epitope Facilitates Efficient T Cell Synapse Formation by Anti-FcRH5/CD3 and Is a Requirement for Myeloma Cell Killing. *Cancer Cell* **2017**, *31*, 383–395. [CrossRef]
51. Dépis, F.; Hatterer, E.; Ballet, R.; Daubeuf, B.; Cons, L.; Glatt, S.; Reith, W.; Kosco-Vilbois, M.; Dean, Y. Characterization of a surrogate murine antibody to model anti-human CD3 therapies. *mAbs* **2013**, *5*, 555–564. [CrossRef] [PubMed]
52. Leo, O.; Foo, M.; Sachs, D.H.; Samelson, L.E.; Bluestone, J.A. Identification of a monoclonal antibody specific for a murine T3 polypeptide. *Proc. Natl. Acad. Sci. USA* **1987**, *84*, 1374–1378. [CrossRef] [PubMed]
53. Tiller, T.; Schuster, I.; Deppe, D.; Siegers, K.; Strohner, R.; Herrmann, T.; Berenguer, M.; Poujol, D.; Stehle, J.; Stark, Y.; et al. A fully synthetic human Fab antibody library based on fixed VH/VL framework pairings with favorable biophysical properties. *mAbs* **2013**, *5*, 445–470. [CrossRef] [PubMed]
54. Valadon, P.; Pérez-Tapia, S.M.; Nelson, R.S.; Guzmán-Bringas, O.U.; Arrieta-Oliva, H.I.; Gómez-Castellano, K.M.; Pohl, M.A.; Almagro, J.C. ALTHEA Gold Libraries™: Antibody libraries for therapeutic antibody discovery. *mAbs* **2019**, *11*, 516–531. [CrossRef] [PubMed]
55. Jain, T.; Sun, T.; Durand, S.; Hall, A.; Houston, N.R.; Nett, J.H.; Sharkey, B.; Bobrowicz, B.; Caffry, I.; Yu, Y.; et al. Biophysical properties of the clinical-stage antibody landscape. *Proc. Natl. Acad. Sci. USA* **2017**, *114*, 944–949. [CrossRef]
56. Kim, D.M.; Yao, X.; Vanam, R.P.; Marlow, M.S. Measuring the effects of macromolecular crowding on antibody function with biolayer interferometry. *mAbs* **2019**, *11*, 1319–1330. [CrossRef]
57. Liu, Y.; Caffry, I.; Wu, J.; Geng, S.B.; Jain, T.; Sun, T.; Reid, F.; Cao, Y.; Estep, P.; Yu, Y.; et al. High-throughput screening for developability during early-stage antibody discovery using self-interaction nanoparticle spectroscopy. *mAbs* **2013**, *6*, 483–492. [CrossRef]
58. Chennamsetty, N.; Voynov, V.; Kayser, V.; Helk, B.; Trout, B.L. Design of therapeutic proteins with enhanced stability. *Proc. Natl. Acad. Sci. USA* **2009**, *106*, 11937–11942. [CrossRef]
59. Kumar, S.; Plotnikov, N.V.; Rouse, J.C.; Singh, S.K. Biopharmaceutical Informatics: Supporting biologic drug development via molecular modelling and informatics. *J. Pharm. Pharmacol.* **2017**, *70*, 595–608. [CrossRef]
60. Voynov, V.; Chennamsetty, N.; Kayser, V.; Helk, B.; Trout, B.L. Predictive tools for stabilization of therapeutic proteins. *mAbs* **2009**, *1*, 580–582. [CrossRef]
61. Glockshuber, R.; Malia, M.; Pfitzinger, I.; Plueckthun, A. A comparison of strategies to stabilize immunoglobulin Fv-fragments. *Biochemistry* **1990**, *29*, 1362–1367. [CrossRef] [PubMed]
62. Jung, S.-H.; Pastan, I.; Lee, B. Design of interchain disulfide bonds in the framework region of the Fv fragment of the monoclonal antibody B3. *Proteins Struct. Funct. Bioinform.* **1994**, *19*, 35–47. [CrossRef] [PubMed]
63. Atwell, S.; Ridgway, J.B.; A Wells, J.; Carter, P. Stable heterodimers from remodeling the domain interface of a homodimer using a phage display library. *J. Mol. Biol.* **1997**, *270*, 26–35. [CrossRef] [PubMed]
64. Schaefer, W.; Regula, J.T.; Bähner, M.; Schanzer, J.; Croasdale, R.; Dürr, H.; Gassner, C.; Georges, G.; Kettenberger, H.; Imhof-Jung, S.; et al. Immunoglobulin domain crossover as a generic approach for the production of bispecific IgG antibodies. *Proc. Natl. Acad. Sci. USA* **2011**, *108*, 11187–11192. [CrossRef]
65. McAleese, F.; Eser, M. RECRUIT-TandAbs®: Harnessing the immune system to kill cancer cells. *Futur. Oncol.* **2012**, *8*, 687–695. [CrossRef]
66. Moore, P.A.; Shah, K.; Yang, Y.; Alderson, R.; Roberts, P.; Long, V.; Liu, D.; Li, J.C.; Burke, S.; Ciccarone, V.; et al. Development of MGD007, a gpA33 x CD3-Bispecific DART Protein for T-Cell Immunotherapy of Metastatic Colorectal Cancer. *Mol. Cancer Ther.* **2018**, *17*, 1761–1772. [CrossRef]
67. Austin, R.; Aaron, W.; Baeuerle, P.; Barath, M.; Jones, A.; Jones, S.D.; Law, C.-L.; Kwant, K.; Lemon, B.; Muchnik, A.; et al. Abstract 1781: HPN536, a T cell-engaging, mesothelin/CD3-specific TriTAC for the treatment of solid tumors. *Immunology* **2018**, *78*, 1781. [CrossRef]
68. Spiess, C.; Zhai, Q.; Carter, P.J. Alternative molecular formats and therapeutic applications for bispecific antibodies. *Mol. Immunol.* **2015**, *67*, 95–106. [CrossRef]
69. Brinkmann, U.; Kontermann, R.E. The making of bispecific antibodies. *mAbs* **2017**, *9*, 182–212. [CrossRef]

70. Ellerman, D. Bispecific T-cell engagers: Towards understanding variables influencing the in vitro potency and tumor selectivity and their modulation to enhance their efficacy and safety. *Methods* **2019**, *154*, 102–117. [CrossRef]
71. Husain, B.; Ellerman, D. Expanding the Boundaries of Biotherapeutics with Bispecific Antibodies. *BioDrugs* **2018**, *32*, 441–464. [CrossRef] [PubMed]
72. Runcie, K.; Budman, D.R.; John, V.; Seetharamu, N. Bi-specific and tri-specific antibodies- the next big thing in solid tumor therapeutics. *Mol. Med.* **2018**, *24*, 1–15. [CrossRef]
73. Shiraiwa, H.; Narita, A.; Kamata-Sakurai, M.; Ishiguro, T.; Sano, Y.; Hironiwa, N.; Tsushima, T.; Segawa, H.; Tsunenari, T.; Ikeda, Y.; et al. Engineering a bispecific antibody with a common light chain: Identification and optimization of an anti-CD3 epsilon and anti-GPC3 bispecific antibody, ERY974. *Methods* **2019**, *154*, 10–20. [CrossRef]
74. Desnoyers, L.R.; Vasiljeva, O.; Richardson, J.H.; Yang, A.; Menendez, E.E.M.; Liang, T.W.; Wong, C.; Bessette, P.H.; Kamath, K.; Moore, S.J.; et al. Tumor-Specific Activation of an EGFR-Targeting Probody Enhances Therapeutic Index. *Sci. Transl. Med.* **2013**, *5*, 207ra144. [CrossRef]
75. Autio, K.A.; Boni, V.; Humphrey, R.W.; Naing, A. Probody Therapeutics: An Emerging Class of Therapies Designed to Enhance On-Target Effects with Reduced Off-Tumor Toxicity for Use in Immuno-Oncology. *Clin. Cancer Res.* **2019**, *26*, 984–989. [CrossRef]
76. Boustany, L.M.; Wong, L.; White, C.W.; Diep, L.; Huang, Y.; Liu, S.; Richardson, J.H.; Kavanaugh, W.M.; Irving, B.A. Abstract A164: EGFR-CD3 bispecific Probody™ therapeutic induces tumor regressions and increases maximum tolerated dose > 60-fold in preclinical studies. In Proceedings of the Abstracts: AACR-NCI-EORTC International Conference: Molecular Targets and Cancer Therapeutics, Philadelphia, PA, USA, 26–30 October 2017; p. A164.
77. Kamata-Sakurai, M.; Narita, Y.; Hori, Y.; Nemoto, T.; Uchikawa, R.; Honda, M.; Hironiwa, N.; Taniguchi, K.; Shida-Kawazoe, M.; Metsugi, S.; et al. Antibody to CD137 activated by extracellular adenosine triphosphate is tumor selective and broadly effective in vivo without systemic immune activation. *Cancer Discov.* **2020**. [CrossRef]
78. Wang, X.; Mathieu, M.; Brezski, R.J. IgG Fc engineering to modulate antibody effector functions. *Protein Cell* **2018**, *9*, 63–73. [CrossRef]
79. Xu, D.; Alegre, M.-L.; Varga, S.S.; Rothermel, A.L.; Collins, A.M.; Pulito, V.L.; Hanna, L.S.; Dolan, K.P.; Parren, P.W.; Bluestone, J.A.; et al. In Vitro Characterization of Five Humanized OKT3 Effector Function Variant Antibodies. *Cell. Immunol.* **2000**, *200*, 16–26. [CrossRef]
80. Dustin, M.L.; Baldari, C.T. The Immune Synapse: Past, Present, and Future. *Adv. Struct. Safety Stud.* **2017**, *1584*, 1–5. [CrossRef]
81. Huppa, J.B.; Gleimer, M.; Sumen, C.; Davis, M.M. Continuous T cell receptor signaling required for synapse maintenance and full effector potential. *Nat. Immunol.* **2003**, *4*, 749–755. [CrossRef]
82. Cartwright, A.N.R.; Griggs, J.; Davis, D.M. The immune synapse clears and excludes molecules above a size threshold. *Nat. Commun.* **2014**, *5*, 5479. [CrossRef] [PubMed]
83. Müller, D.; Karle, A.; Meissburger, B.; Höfig, I.; Stork, R.; Kontermann, R.E. Improved Pharmacokinetics of Recombinant Bispecific Antibody Molecules by Fusion to Human Serum Albumin. *J. Biol. Chem.* **2007**, *282*, 12650–12660. [CrossRef] [PubMed]
84. Stork, R.; Campigna, E.; Robert, B.; Müller, D.; Kontermann, R.E. Biodistribution of a Bispecific Single-chain Diabody and Its Half-life Extended Derivatives. *J. Biol. Chem.* **2009**, *284*, 25612–25619. [CrossRef] [PubMed]

Publisher's Note: MDPI stays neutral with regard to jurisdictional claims in published maps and institutional affiliations.

Article

Characterization of Co-Formulated High-Concentration Broadly Neutralizing Anti-HIV-1 Monoclonal Antibodies for Subcutaneous Administration

Vaneet K. Sharma [1,†], Bijay Misra [2,†], Kevin T. McManus [3], Sreenivas Avula [1], Kaliappanadar Nellaiappan [2], Marina Caskey [4], Jill Horowitz [4], Michel C. Nussenzweig [4,5], Michael S. Seaman [3], Indu Javeri [2] and Antu K. Dey [1,*]

1 IAVI, 125 Broad Street, New York, NY 10004, USA; VSharma@iavi.org (V.K.S.); SAvula@iavi.org (S.A.)
2 CuriRx, Inc., 205 Lowell Street, Wilmington, MA 01887, USA; bjmisra@gmail.com (B.M.); nellaiappan@curirx.com (K.N.); Ijaveri@curirx.com (I.J.)
3 Center for Virology and Vaccine Research, Beth Israel Deaconess Medical Center, Boston, MA 02215, USA; kmcmanu4@bidmc.harvard.edu (K.T.M.); mseaman@bidmc.harvard.edu (M.S.S.)
4 Laboratory of Molecular Immunology, The Rockefeller University, New York, NY 10065, USA; mcaskey@mail.rockefeller.edu (M.C.); jhorowitz@mail.rockefeller.edu (J.H.); nussen@mail.rockefeller.edu (M.C.N.)
5 Howard Hughes Medical Institute, The Rockefeller University, New York, NY 10065, USA
* Correspondence: adey@iavi.org
† Both authors contributed equally to the work.

Received: 26 June 2020; Accepted: 23 July 2020; Published: 29 July 2020

Abstract: The discovery of numerous potent and broad neutralizing antibodies (bNAbs) against Human Immunodeficiency Virus type 1 (HIV-1) envelope glycoprotein has invigorated the potential of using them as an effective preventative and therapeutic agent. The majority of the anti-HIV-1 antibodies, currently under clinical investigation, are formulated singly for intra-venous (IV) infusion. However, due to the high degree of genetic variability in the case of HIV-1, a single broad neutralizing antibody will likely not be sufficient to protect against the broad range of viral isolates. To that end, delivery of two or more co-formulated bnAbs against HIV-1 in a single subcutaneous (SC) injection is highly desired. We, therefore, co-formulated two anti-HIV bnAbs, 3BNC117-LS and 10-1074-LS, to a total concentration of 150 mg/mL for SC administration and analyzed them using a panel of analytical techniques. Chromatographic based methods, such as RP-HPLC, CEX-HPLC, SEC-HPLC, were developed to ensure separation and detection of each antibody in the co-formulated sample. In addition, we used a panel of diverse pseudoviruses to detect the functionality of individual antibodies in the co-formulation. We also used these methods to test the stability of the co-formulated antibodies and believe that such an approach can support future efforts towards the formulation and characterization of multiple high-concentration antibodies for SC delivery.

Keywords: HIV/AIDS; co-formulation; high concentration; analytical characterization; antibody (s)

---

1. Introduction

The number of approved monoclonal antibodies (mAbs) for therapy against various cardiovascular, cancer, respiratory, hematology, and autoimmune diseases is continuously on the rise [1]. In addition to therapy against non-infectious diseases, monoclonal antibodies are also increasingly seen as potent prophylactic and therapeutic agents against several infectious pathogens [2–5], particularly those against which effective vaccines do not exist or are under arduous development. To date, over a hundred antibodies have been approved by various regulatory authorities; the majority of these antibody products are typically administered by intravenous (IV) infusion. IV administration, although a well-established route, is challenging to patients as well as to healthcare professionals. Subcutaneous (SC) administration, on the other hand, is increasingly becoming a clear patient preference due to time savings and potential for self-administration, including possibilities for healthcare professionals of administrating during home visits to patients [6,7].

The use of monoclonal antibodies as prophylactic and therapeutic options is particularly attractive against Human Immunodeficiency Virus type 1 (HIV-1) [8,9], a viral pathogen for which the development timeline for a prophylactic vaccine is uncertain [10–12]. Therefore, protection using passive administration of broadly neutralizing antibodies (bNAbs) against HIV-1 is being evaluated through multiple human clinical studies to test the validity of the approach. Broadly neutralizing (monoclonal) antibodies (bNAbs) such as VRC01 [13,14], 10-1074 [15]/10-1074-LS [16], 3BNC117 [17]/3BNC117-LS [18], VRC07-523-LS [19], PGT121 [20,21], and PGDM1400 [21] or their combinations are currently under investigation in multiple clinical trials. Recent studies by Bar-On et al. [22] and Mendoza et al. [23] showed that the combination of two bNAbs, 3BNC117 (directed to CD4-binding site epitope on HIV-1 surface envelope glycoprotein) [17,24] and 10-1074 (directed to V3-glycan epitope on HIV-1 surface envelope glycoprotein) [15,24], delivered by the intravenous (IV) route was well-tolerated and effective in maintaining virus suppression for extended periods in individuals harboring HIV-1 strains sensitive to the antibodies. These clinical studies, with safety, pharmacokinetics and viral load re-bound or decay as endpoints, have primarily used antibodies formulated for IV infusion. Moving forward, to overcome the high cost and burden of intra-venous administration, the high-concentration formulation of both antibodies (here referred to as co-formulation) for sub-cutaneous (SC) administration is planned. However, co-formulating two (or more) antibodies at high concentration is not only challenging due to the requirement to maintain their optimal quality attributes, low viscosity and stability in the chosen formulation condition but also in developing analytical methods that allow separation of individual antibodies to characterize their quality attributes and measure their individual and total stability [25]. Recently, Cao et al. reported the characterization of antibody charge variants and the development of "release" assays for co-formulated antibodies [26]. In another study, Patel et al. investigated the formulation of two anti-HIV bNAbs and through a series of analytical tools, including the mass spectrometry-based multi attribute method (MAM), the authors highlight the analytical challenges in the characterization of co-formulated antibodies [27].

Here, we describe the formulation of two high-concentration bnAbs, 3BNC117-LS and 10-1074-LS, to a final concentration of 150 mg/mL and characterize them through a panel of analytical methods to evaluate the suitability of the methods for future cGMP testing of the co-formulated drug product. Additionally, we show that the chromatography-based separation methods (RP-HPLC, SE-HPLC and IEX-HPLC) and virus-based neutralization assay are optimal to study each antibody in the co-formulated milieu and can potentially be used for "release" and "stability" testing of these materials.

2. Materials and Methods

2.1. Materials

2.1.1. Monoclonal Antibodies

3BNC117 is a monoclonal antibody of the IgG1κ isotype that specifically binds to the CD4 binding site (CD4bs) within HIV-1 envelope gp120. The bnAbs 10-1074 is of the IgG1λ isotype that specifically targets the V3 glycan supersite within HIV-1 envelope gp120. Both fully human parental monoclonal antibodies, 10-1074 and 3BNC117, were LS-modified, two amino acid substitutions, Methionine (M) to Leucine (L) at Fc position 428 (M428L) and Asparagine (N) to Serine (S) at Fc position 434 (N434S), to enhance the antibody binding affinity to the neonatal Fc receptor (FcRn) and prolong their half-life in mammals without impacting the antibody binding domain or its interaction with antigens [28,29]. The LS-modified monoclonal antibodies (MAbs) are referred to here as 3BNC117-LS and 10-1074-LS. The 3BNC117-LS and 10-1074-LS mAbs were produced at Celldex Therapeutics (Fall River, MA, USA).

Both antibodies were expressed via stable Chinese hamster ovary (CHO) cell line clones in a serum-free medium in a batch bioreactor using standard mammalian cell culture techniques. The harvested clarified supernatant was then used to purify the mAbs using a series of chromatographic steps that included MabSelect Sure, Sartobind Q, and SP Sepharose cation exchange column chromatography's. The SP Sepharose eluate was nano-filtered using Virosart HG filtration and concentrated to 150 mg/mL concentration by UFDF (Ultra-filtration Dia-filtration).

The 3BNC-117-LS monoclonal antibody concentrated to 150 mg/mL was formulated in a buffer containing 10 mM Methionine, 250 mM Trehalose, 0.05% Polysorbate 20, pH 5.2. The 10-1074-LS monoclonal antibody concentrated to 150 mg/mL was formulated in a buffer containing 5 mM Histidine, 250 mM Trehalose, 10 mM Methionine, 5 mM Sodium Acetate, 0.05% Polysorbate 20, pH 5.5.

For this study, 3BNC117-LS and 10-1074-LS, were co-formulated (1:1) by mixing at ambient conditions, and the buffer was exchanged such that the final formulation buffer was 5 mM Histidine, 250 mM Trehalose, 10 mM Methionine, 5 mM Sodium Acetate, 0.05% Polysorbate 20, pH 5.5.

2.1.2. Reagents

The hybridoma-based monoclonal anti-idiotype antibodies, used in the ELISA, were produced at Duke Human Vaccine Institute (Durham, NC, USA). The hybridomas were created by immunizing BALB/C mice with either 10-1074 Fab fragment or 3BNC117 Fab fragment. The generated anti-idiotype antibodies were chromatographically purified and concentrated to ~7 mg/mL in 1 × PBS pH 7.2, 0.22 µm filtered and stored at 4 °C until further use. The USP grade Histidine, Methionine, Polysorbate 20 were purchased from JT Baker Chemicals (Phillipsburg, NJ, USA) and Trehalose was purchased from Pfanstiehl, Inc. (Waukegan, IL, USA). All solutions were stored at 4 °C until used.

2.2. Methods

2.2.1. Reverse Phase High-Performance Liquid Chromatography (RP-HPLC)

RP-HPLC separation was performed on Agilent 1260 Infinity Quaternary LC coupled to a diode array detector (DAD). Best peak resolution was demonstrated using Agilent AdvanceBio RP-mAb Diphenyl, 2.1 × 100 mm column, 0.5 mL/min flow rate, with a column temperature of 60 °C and a step wise gradient (3 min washing at 35% B followed by 35% B to 39% B over 16 min). The eluted peaks were detected at 280 nm.

2.2.2. Ion Exchange (IEX)—HPLC

IEX-HPLC was performed on the Agilent 1260 Infinity Quaternary LC system equipped with a diode array detector (DAD) and coupled to ProPac WCX-10, 250 × 4 mm column (Thermo Scientific, Sunnyvale, CA, USA) maintained at 30 °C. Mobile phase A consisted of 20 mM Acetate, pH 5.2, while mobile phase B was 20 mM Acetate, 300 mM sodium chloride, pH 5.2. The pHs of both mobile phases was adjusted using 0.1 M NaOH solution. The flow rate was 0.7 mL/min and salt gradient separation, 50% to 100% B in 35 min, was performed. Peak detection was carried out at 280 nm and the peaks were integrated and percentage peak areas of each peak (as well as charge variants i.e., acidic/basic species) calculated corresponding to each mAbs.

2.2.3. Size-Exclusion High-Performance Liquid Chromatography (SE-HPLC)

SE-HPLC was performed on the Agilent 1260 Infinity Quaternary LC system equipped with a diode array detector (DAD) and coupled to TSKgel G3000SWXL, 5 µm, 7.8 mm × 30 cm column maintained at 30 °C. The mobile phase used was 10 mM histidine, 50 mM Arginine, 100 mM sodium sulfate, pH 6.0. The flow rate used was 1 mL/min. The eluted main and High-Molecular Weight (HMW) peaks were detected at 280 nm.

2.2.4. Enzyme-Linked Immunosorbent Assay (ELISA)

A sandwich ELISA was performed using 96-well Maxisorp plates coated over-night at 2–8 °C with 1 μg/mL of an anti-idiotypic antibody that specifically recognizes 3BNC117-LS (anti-ID monoclonal antibody) or 1 μg/mL of an anti-idiotypic antibody that specifically recognizes 10-1074-LS (anti-ID monoclonal antibody). After washing, plates were blocked with 200 μL Protein free blocking solution at 25 °C for 2 h at 200 RPM. Co-formulated antibody samples, quality controls and reference standards were added and incubated at room temperature. Subsequently, the plate was washed and 100 μL of 1:10,000 diluted peroxidase-conjugated AffiniPure F (ab′) 2 Fragment Goat anti-Human IgG Fcγ Fragment specific (Jackson Immuno Research, West Grove, PA, USA) was added. The plate was incubated at room temperature for 60 ± 10 min at 200 RPM. The plate was washed, and the wells were incubated with 100 μL of SureBlue TMB substrate (Fisher Scientific, Somerset, NJ, USA) to develop the chromogenic signal (10 min at room temperature at 200 RPM). The reaction was stopped with the addition of 100 μL of 1% hydrochloric acid. The absorbance was measured at 450 nm using the Molecular Devices plate reader fitted with Softmax Pro software (Molecular Devices LLC, Sunnyvale, CA, USA). Titration curves for the reference standard and each test sample were created using 4-parameter logistic curve fitting to calculate EC50 values using GraphPad Prism software (version 7).

2.2.5. Virus Neutralization Assays

The virus neutralization assay was evaluated using a luciferase-based assay in TZM-bl cells, as previously described [30,31]. Briefly, antibody samples were tested using a starting concentration of 25 μg/mL with 5-fold serial dilutions against the panel of HIV-1 Env pseudoviruses. The selected panel of HIV-1 Env pseudoviruses were either 3BNC117 sensitive/10-1074 resistant (n = 10) or 3BNC117 resistant/10-1074 sensitive (n = 10). The IC50 and IC80 titers were calculated as the mAb concentration that yielded a 50% or 80% reduction in relative luminescence units (RLU), respectively, compared to the virus control wells after the subtraction of cell control RLUs. All assays were performed in a laboratory compliant with Good Clinical Laboratory Practice (GCLP) procedures.

2.2.6. FlowCAM® Imaging

FlowCAM® is an imaging particle analysis system that we used for imaging and analyzing particles, in the subvisible range, using flow microscopy. The FlowCAM® instrument (Fluid Imaging Technologies, Scarborough, ME, USA) was focused with 10 μm polystyrene beads at 3000/mL National Institute of Standards and Technology (NIST) standard. The samples were diluted by 4-fold by taking 200 μL of the sample in the corresponding buffer to a total volume of 0.8 mL, samples were analyzed at 0.08 mL/min through a 100 μm × 2 mm flow cell, and images of the particles were taken with a 10× optics system. Flash duration was set to 35.50 ms, and Camera Gain was set to 0. Visual-Spreadsheet software version 3.4.8 (DKSH Japan K.K., Tokyo, Japan) was used for data analysis.

2.2.7. Osmolality

Osmolality, a measurement of the total number of solutes in a liquid solution expressed in osmoles of solute particles per kilogram of solvent (mOsm/Kg), was measured in the antibody formulations using the industry-preferred freezing point depression method. The osmolality measurements were made using a Model 3340 single-sample freezing-point micro-osmometer (Advanced Instruments, Norwood, MA, USA), equipped with a 20 μL Ease Eject™ Sampler (Parts No. 3M0825 and 3M0828). The unit of measurement used was milliosmoles of the solute per 1 kg of pure solvent, expressed as mOsm/kg. The instrument was calibrated with 50 mOsm/kg (3MA005) and 850 mOsm/kg (3MA085) calibration standards and verified with a 290 mOsm/kg Clinitrol® Reference Solution (3MA029) prior to each analysis.

2.2.8. Dynamic Light Scattering (DLS)

Dynamic Light Scattering (DLS), which uses time-dependent fluctuations in the intensity of the scattered light to determine the effective size of a particle in nm range, was used to measure the particle size distribution in the antibody formulations. Dynamic light scattering was carried out at 25 °C, with a Malvern Zetasizer Nano Series instrument using a 633 nm/100 mW laser and a 90° detection angle. Particle size distribution (hydrodynamic diameter) by % intensity and % volume was determined along with the polydispersity index (PDI).

3. Results

Parental anti-HIV antibodies, 10-1074 and 3BNC117, were individually formulated at 20 mg/mL for IV administration [15,17]. Based on initial pK data from phase 1 clinical studies, the parental antibodies were LS modified (as described in the Materials section) to extend serum half-life. Thereafter, as a first step towards formulation to aid subcutaneous administration, the antibodies were concentrated 7.5-fold in a new formulation buffer with optimal viscosity to enable drug injection volumes of 2 mL. This resulted in 3BNC117-LS and 10-1074-LS as individually formulated bnAbs, at 150 mg/mL in respective buffers (described in the Materials and Methods section), for subsequent co-formulation studies.

These high-concentration individually formulated antibodies were extensively characterized using a wide range of analytical methods i.e., ELISA, SE-HPLC, RP-HPLC, IEX-HPLC, capillary isoelectric focusing (cIEF), CE-SDS (reduced and non-reduced), Sialic Acid analysis, intrinsic Tryptophan fluorescence spectroscopy, Isoquant analysis, dynamic light scattering (DLS), far and near UV circular dichroism (CD), differential scanning calorimetry (DSC), second derivative UV spectroscopy, N-terminal amino acid sequencing, HILIC based glycan profiling, liquid chromatography coupled with mass spectrometry (LC/MS), and peptide mapping by liquid chromatography coupled with tandem mass spectrometry (LC-MS/MS) (data not shown). In addition, both high-concentration bNAbs, in their respective formulation conditions, were found to be stable at 2–8 °C for ≥24 months (data not shown).

To address the need to co-formulate both the antibodies as a single drug product at a combined final concentration of 150 mg/mL for subcutaneous administration in clinical studies, a 1:1 mixture of both bNAbs (each at 75 mg/mL) was formulated in 5 mM Histidine, 250 mM Trehalose, 10 mM Methionine, 5 mM Sodium Acetate, 0.05% Polysorbate 20, pH 5.5 buffer and characterized using a series of methods to test for (positive) identity, purity, product qualities, and functionality.

3.1. Chromatographic Separation of Co-Formulated Monoclonal Antibodies

During analytical development, the aim was to select appropriate and optimal chromatographic techniques that could separate the two antibodies in their current co-formulation. Reverse phase high-performance liquid chromatography (RP-HPLC), ion exchange liquid chromatography (IEX), and size exclusion chromatography (SEC) were evaluated and found to achieve this separation goal. The separation efficiencies of each of these methods were challenged by the fact that both antibodies, 3BNC-117-LS and 10-1074-LS, are of the IgG1 subclass with similar molecular size and three-dimensional structure, and therefore significant method development and optimization of the chromatographic methods were necessary to achieve the desired separation goals.

3.2. Reverse Phase High-Performance Liquid Chromatography (RP-HPLC)

Reverse phase liquid chromatography (RP-HPLC), due to the denaturing effect of the low pH and high organic solvent mobile phase, was expected to separate and quantify the two bNAbs, in the co-formulated milieu, based on their differences in the relative hydrophobicity. Initial assessment was performed on the Agilent 1260 Infinity quaternary LC coupled to a DAD detector using two columns: AdvanceBio RP-mAb Diphenyl, 2.1 × 100 mm, 3.5 μm (Agilent) and Accucore 150-C4 2.6 μm, 100 × 2.1 mm (Thermo). These columns represent two different stationary phase chemistries, diphenyl offers alternative selectivity and Accucore C-4 wide pore (150 Å) offers lower hydrophobic retention. A generic method development strategy was followed using 0.1% TFA in acetonitrile as mobile phase B, 0.5 mL/minute flow rate. Since, the individually formulated antibodies did elute at 30–40% acetonitrile at elevated temperature (60 °C column temperature) (data not shown), the initial gradient conditions for developing the method was set at 34% to 41% mobile phase B for 7 min. As part of the method optimization, different chromatographic conditions were tested to increase peak resolution: increasing temperature (45, 60, 70, and 80 °C), different mobile phases or organic modifiers in acetonitrile (methanol as mobile phase B or methanol/ IPA (5% *v*/*v*) as organic modifier in acetonitrile mobile phase), and different gradient conditions (34% B to 38% B for 16 min and 35% B to 39% B for 16 min). Finally, a method with 34% to 41% mobile phase B, and a column temperature of 60 °C, was selected that resulted in two separate peaks, corresponding to each monoclonal antibody (Figure 1A). This RP-HPLC method was then tested for linearity, precision, accuracy, and specificity parameters. Linearity was evaluated for the total peak area of 10-1074-LS and 3BNC117-LS in a co-formulated sample by calculation of a regression line using the least squares method. (Figure 1B). Linearity for the 3BNC117-LS specific area (Figure 1C) and 10-1074-LS specific area (Figure 1D) were also calculated; the R^2 obtained was 1 for both analyses. Precision, for intra and inter day variability, was assessed by testing the repeatability of the target concentration of 2.5 µg (for each antibody) six times. The intra-assay precision for the total peak area ranged from 0.1% to 0.3% and the inter-assay precision was 0.3% for 3 experiments on separate days (days 1, 2, and 3) (Supplementary Table S1A). The intra- and inter-precisions for the individual peak areas, 10-1074-LS peak area and 3BNC117-LS peak area, were also similar. The intra- and inter-precisions for the 10-1074-LS peak area were ≤0.4% and ≤0.3%, respectively (Supplementary Table S1B). The intra- and inter-precisions for the 3BNC117-LS peak area were ≤0.2% and ≤0.2%, respectively (Supplementary Table S1C). Accuracy was tested by percentage recoveries of the mean of three determinations of six different concentrations (1 to 8 µg column load). Based on the percent recoveries, we concluded that the RP-HPLC method accuracy was within a variation of ≤2% relative standard deviation (RSD) (Supplementary Table S1D). These results indicate that this RP-HPLC method is suitable and, after appropriate method validation, can be used for future testing of the 3BNC117-LS + 10-1074-LS co-formulated drug product.

3.3. *Ion Exchange High-Performance Liquid Chromatography (IEX-HPLC)*

To allow the characterization of charge heterogeneity and high-resolution separation of each antibody (in the co-formulated sample), it was expected that the ion exchange (IEX) chromatography can separate the two bNAbs based on their charge differences. IEX chromatography is a non-denaturing technique and among the different IEX modes, since cation-exchange chromatography (CEX) is the preferred approach for characterizing antibody charge variants [32,33], it was chosen. The CEX method was developed on the Agilent 1260 Infinity quaternary LC system equipped with a solvent delivery pump, an autosampler, and a diode array detector (DAD). A ProPac WCX-10, 250 × 4 mm column (Thermo Scientific, Sunnyvale, CA, USA) was used for the method development. A "classical" salt gradient separation (50% to 100% B in 35 min) was performed using mobile phase A, composed of 20 mM Acetate buffer, pH 5.2, and mobile phase B, composed of 20 mM Acetate buffer containing 300 mM sodium chloride, pH 5.2. The flow rate was set to 0.7 mL/min and column temperature was maintained at 30 °C. Peak detection was carried out at 280 nm and after integration of peaks, the percentage peak areas of each peak (as well as charge variants i.e., main, acidic, basic peak) corresponding to each monoclonal antibody were calculated (Figure 2A). This weak cation exchange (CEX) chromatography method was used to perform qualitative and quantitative analysis of the charge variants for each of the separated antibodies. The optimized CEX-HPLC method was further tested for linearity, precision, accuracy, and specificity parameters. Linearity was evaluated for main peaks, pre-main peaks, post main peaks, and total peak areas of 10-1074-LS and 3BNC117-LS in a co-formulated sample by calculation of a regression line using the least squares method (Figure 2B–D). Precision, for intra- and inter-assay variability, of the total peak area was assessed by testing the repeatability of the target concentration of 100 µg (total) six times. The charge variants for both

10-1074-LS and 3BNC117-LS were within 2.2% for the total peak area with intra-assay precision within 1.1–2.2% and inter-assay precision within <2.2% (Supplementary Table S2A). The variability of the 10-1074-LS specific peak was similar, with intra-assay precision ≤ 2.1% and inter-assay precision ≤ 1.8% (Supplementary Table S2B). The variability of the 3BNC117-LS specific peak was slightly higher, although similar, with intra-assay precision ≤ 2.6% and inter-assay precision ≤ 2.7% (Supplementary Table S2C). The accuracy was tested by percentage recoveries of the mean of three determinations of six different concentrations (50 to 300 μg column load); the method accuracy was observed to be ≤2% relative standard deviation (RSD) (Supplementary Table S2D). Based on these results, this CEX-HPLC method is suitable and after appropriate method validation can be used for future testing of the 3BNC117-LS + 10-1074-LS co-formulated drug product.

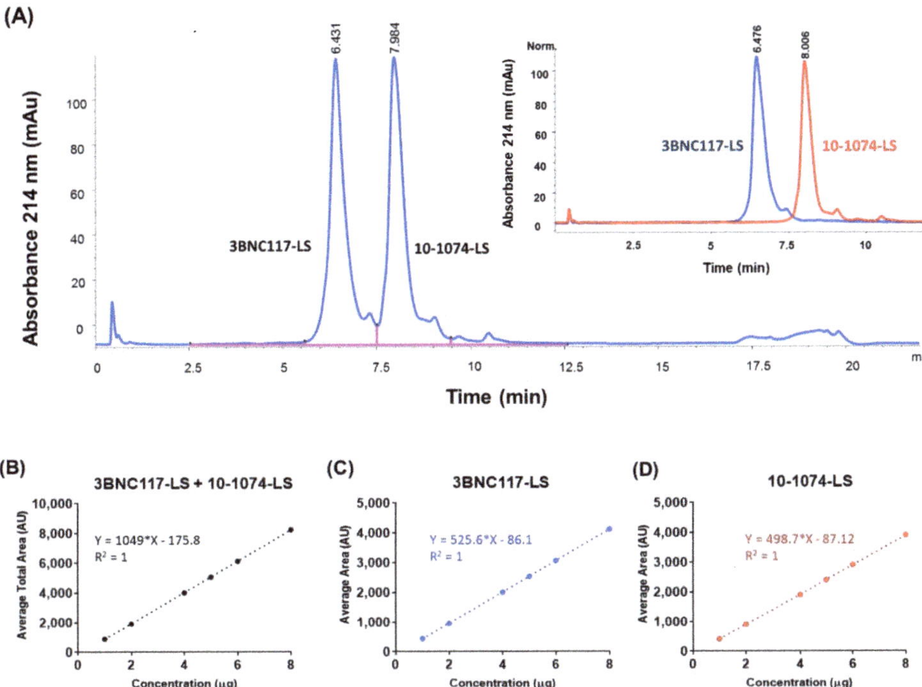

Figure 1. Reverse phase chromatogram with UV absorbance at 214 nm showing separation of two antibodies in the 1:1 co-formulated sample (total 150 mg/mL). (**A**) Peaks separated by reversed phase HPLC corresponding to the two antibodies are labeled. Inset shows the overlapped reverse phase chromatograms from the two-separate RP-HPLC run corresponding to the two antibodies, each at 150 mg/mL. The chromatography conditions were the same for the co-formulated and the individual antibody samples. Linearity analysis of concentration (in μg; x-axis) dependent increase in area under the curve (in Absorbance Units, AU; y-axis) for (**B**) total area (3BNC117-LS + 10-1074-LS), (**C**) 3BNC117-LS specific area, and (**D**) 10-1074-LS specific area.

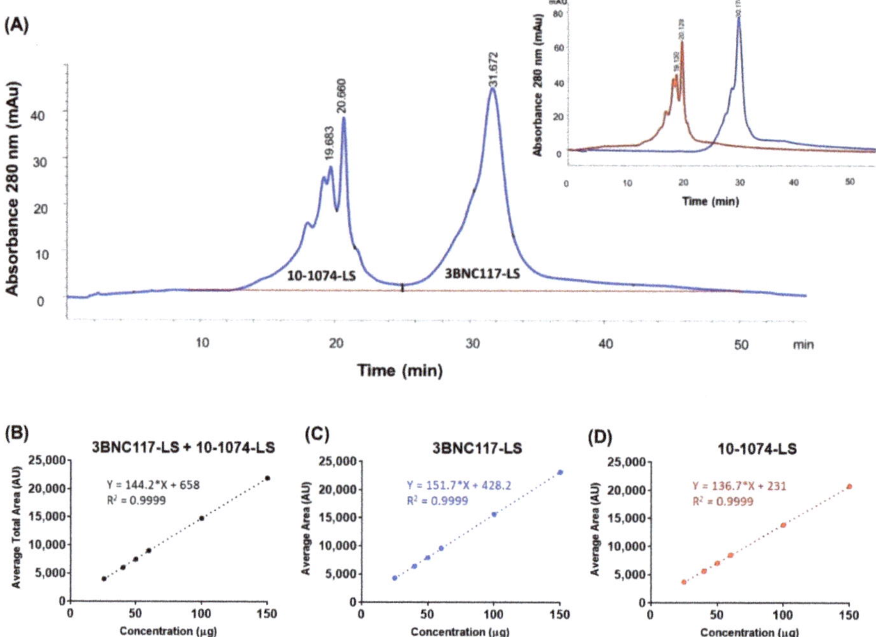

Figure 2. (**A**) Cation exchange chromatogram with UV absorbance at 280 nm showing separation of two antibodies in the 1:1 co-formulated sample (total 150 mg/mL). Peaks for each of the antibody are labeled. Inset shows the overlapped cation exchange chromatograms from the two separate chromatography runs corresponding to the two antibodies, each at 150 mg/mL. The chromatography conditions were the same for the co-formulated and the individual antibody samples. Linearity analysis of concentration (in µg; x-axis) dependent increase in area under the curve (in Absorbance Units, AU; y-axis) for (**B**) total area (3BNC117-LS + 10-1074-LS), (**C**) 3BNC117-LS specific area, and (**D**) 10-1074-LS specific area.

3.4. Size-Exclusion High-Performance Liquid Chromatography (SE-HPLC)

To separate the two antibodies (in the co-formulated sample) based on their molecular size and achieve separation through differential exclusion, we used Size Exclusion HPLC (SE-HPLC). SE-HPLC is widely used for determining the antibody purity, through the determination of percent monomer (and assessments of % HMW, High Molecular Weight, and % LMW, Low Molecular Weight, species), and therefore we were conscious of the possible limitation of the method to fully resolve the two similarly sized monoclonal antibodies in the co-formulated sample. When we evaluated two mobile phases (100 mM sodium acetate (pH 6.0) and 100 mM sodium sulfate (pH 6.0)) for the resolution of the antibodies in the co-formulated sample, we found that both phases resulted in one broad peak with no resolution of the two antibodies (data not shown). However, when we changed the mobile phase to 10 mM histidine, 50 mM arginine, 100 mM sodium sulfate, pH 6.0, a slight separation between the two peaks was observed (Figure 3A). When the salt (sodium sulfate) concentration was gradually increased from 100 to 550 mM sodium sulfate, to probe the effect of increasing salt on the separation of the two peaks (10-1074-LS and 3BNC117-LS peaks), we observed separation with greater resolution, despite the broadening of the late eluting 10-1074-LS peak. From the outset, since the intent of the SE-HPLC was not to resolve the two mAbs but to detect the levels of HMW (and LMW) species in the co-formulated sample and quantify aggregate levels (at the time of product release and during long-term storage) to ensure a means for measurement of percent monomeric antibody in the co-formulated milieu, this SE-HPLC method was accepted to be appropriate for use and tested further for linearity, precision, accuracy, and specificity parameters. Linearity was verified in a range of co-formulated samples with a R^2 of >0.99 (Figure 3B–D). The intra- and inter-precision at the target column load of 100 µg (total) for the main peak (monomer) area was within ≤0.2% (Supplementary Table S3A). The intra- and inter-precision at the target column load of 100 µg (total) for the % HMW peak area was within ≤0.7% and ≤4.3% (Supplementary Table S3B). However, the intra- and inter-precision at the target column load of 100 µg (total) for the % LMW peak area was higher, within ≤18.4% and ≤14% (Supplementary Table S3C); this higher % RSD was due to lower signal levels (lower levels of LMW), closer or below limit of quantification (LOQ). Accuracy was tested by percentage recoveries of the mean of three determinations of six different concentrations precisely prepared (50 to 300 µg column load); the method accuracy was observed to be within 2% RSD (Supplementary Table S3D). These results indicate that the SE-HPLC method is suitable and after appropriate method validation can be used for future testing of the 3BNC117-LS + 10-1074-LS co-formulated drug product.

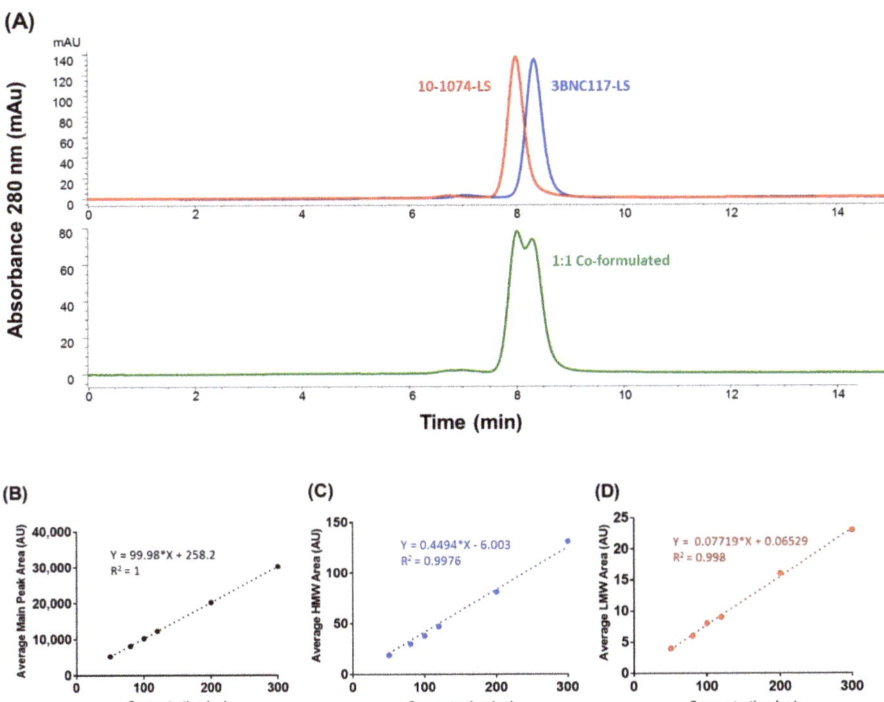

Figure 3. (**A**) Size exclusion chromatography profiles of 150 mg/mL for 10-1074-LS and 3BNC117-LS antibodies, individually formulated (top) and after co-formulation (1:1, 75 mg/mL each) (bottom) on the TSKgel column. Linearity analysis of concentration (in µg; x-axis) dependent increase in area under the curve (in Absorbance Units, AU; y-axis) for (**B**) average main peak (monomer) area (3BNC117-LS + 10-1074-LS monomer), (**C**) average High-Molecular Weight (HMW) peak area, and (**D**) average Low Molecular Weight peak area.

3.5. Positive Identification of Individual Antibody in the Co-Formulation Sample Using Anti-ID (Idiotype) Based ELISA

Since wildtype gp120-based ELISA would not be successful in differentiating the binding and the identity of the two antibodies, when present in a co-formulated sample, we generated a 10-1074 idiotype-specific antibody and a 3BNC117 idiotype-specific antibody to serve as reagents in a new ELISA that would utilize each antibody's identity based on their unique idiotype (ID). This format would provide a means for measuring the identity of an individual antibody in the co-formulated sample. After a series of optimization experiments, the anti-ID ELISA was successful to identify and differentiate both antibodies as well as detect their identity in the co-formulated sample (Figure 4). The anti-ID ELISA was tested for precision and accuracy (data not shown) and the overall variability, particularly inter-assay, was well within the 30–40% RSD, seen in bioassays (data not shown). These results indicate that the anti-ID based ELISA can be used for future testing of identity of the 3BNC117-LS + 10-1074-LS co-formulated drug product, after appropriate method validation.

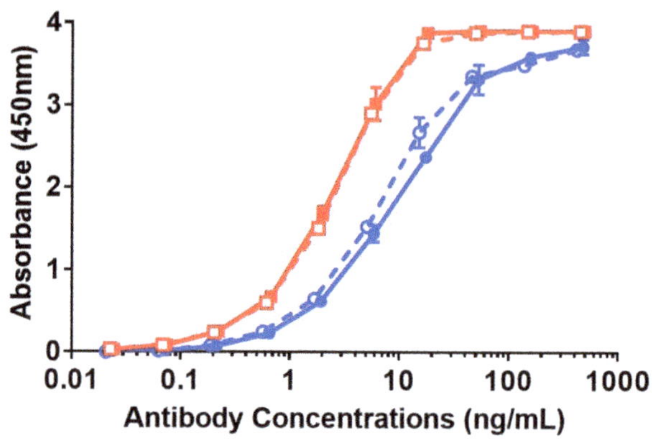

Figure 4. Test of identity of individual antibodies (3BNC117-LS and 10-1074-LS) in co-formulated sample (of 150 mg/mL total concentration) using anti-idiotypic (anti-ID) antibodies. Individual antibodies, 3BNC117-LS and 10-1074-LS at 150 mg/mL, were used as a reference standard. Open circle, dotted line—3BNC117-LS (150 mg/mL, reference standard), filled circle, filled line—3BNC117-LS (at 75 mg/mL in co-formulated sample), open square, dotted line—10-1074-LS (150 mg/mL, reference standard), filled square, filled line—10-1074-LS (at 75 mg/mL in co-formulated sample).

3.6. *Potency Testing of Individual Antibody in the Co-Formulation Sample Using a Virus Neutralization Assay*

A traditional HIV-1 pseudovirus neutralization assay was used to evaluate the functional activity or potency of the individual antibodies in the co-formulated sample. To do so, two panels of pseudoviruses (total $n = 20$) were selected: one panel ($n = 10$) for 3BNC117 and another ($n = 10$) for 10-1074. To test for 3BNC117 potency/functional activity, the panel involved ten 3BNC117 sensitive/10-1074 resistant viruses; to test for 10-1074 potency/functional activity, the panel involved ten 3BNC117 resistant/10-1074 sensitive viruses. Viruses were selected for having low/medium to high sensitivity to a single antibody based on historical data (data not shown). MuLV (Murine Leukemia Virus), a non-relevant virus, was used as a negative control and was not neutralized by either of the two control antibodies (data not shown). In comparison to the individual antibody (3BNC117 or 10-1074), used as control, the two co-formulated antibodies were potent and demonstrated their specific neutralization activity in their respective panels (Table 1A–D). Out of the 10 viruses sensitive to 3BNC117, 0013095-2.11 is not highly sensitive to 3BNC117; therefore, not only is the IC80 >25 µg/mL, the IC80 of the co-formulated 3BNC117-LS and 10-1074-LS is also higher when compared to IC80s for other viruses for the co-formulated product. (Table 1A). Despite this one virus, which could be replaced by another virus in future, based on these results, the pseudovirus neutralization assay using the defined panel of viruses can be used for future testing of the 3BNC117-LS + 10-1074-LS co-formulated drug product.

Table 1. Neutralization activity of co-formulated 3BNC117-LS and 10-1074-LS (total 150 mg/mL) using 2 panels of pseudoviruses in TZM-bl cells. One panel (**A**,**B**) is used to test potency/functional activity of 3BNC117 (3BNC117 sensitive/10-1074 resistant viruses (n = 10)) and the other panel (**C**,**D**) is used to test potency/functional activity of 10-1074 (10-1074 sensitive viruses/3BNC117 resistant (n = 10)). The pseudovirus strains are indicated at the top of the table and IC50 and IC80 values (in μg/mL) for each of the samples, against those viruses, are reported. Individual antibodies, 3BNC117 and 10-1074, are used as controls for each panel. LS—Leucine-Serine substitution.

(A)										
Samples	**ZM249M.PL1**		**Q461.e2**		**0013095-2.11**		**62357.14.D3.4589**		**ZM53M.PB12**	
3BNC117.LS + 10-1074.LS DP	IC50	IC80	IC50	IC80	IC50	IC80	IC50	IC80	IC50	IC80
	0.042	0.15	0.042	0.153	1.161	15.162	0.043	0.15	0.153	0.568
3BNC117.LS (control)	0.037	0.13	0.039	0.143	1.396	>25	0.036	0.17	0.214	0.796
10-1074.LS (control)	>25	>25	>25	>25	>25	>25	>25	>25	>25	>25

(B)										
Samples	**C2101.c01**		**C4118.c09**		**THRO4156.18**		**415.v1.c1**		**CNE5**	
3BNC117.LS + 10-1074.LS DP	IC50	IC80	IC50	IC80	IC50	IC80	IC50	IC80	IC50	IC80
	0.029	0.135	0.034	0.162	1.498	8.209	0.05	0.115	0.193	0.898
3BNC117.LS (control)	0.044	0.15	0.051	0.183	1.939	9.815	0.048	0.142	0.193	0.911
10-1074.LS (control)	>25	>25	>25	>25	>25	>25	>25	>25	>25	>25

(C)										
Samples	**1394C9_G1 (Rev-)**		**ZM247v1 (Rev-)**		**Du422.1**		**6631.v3.c10**		**377.v4.c9**	
3BNC117.LS + 10-1074.LS DP	IC50	IC80	IC50	IC80	IC50	IC80	IC50	IC80	IC50	IC80
	0.02	0.077	0.02	0.099	0.032	0.114	0.157	0.796	0.418	1.397
3BNC117.LS (control)	>25	>25	>25	>25	>25	>25	>25	>25	>25	>25
10-1074.LS (control)	0.033	0.119	0.036	0.162	0.045	0.161	0.189	0.968	0.433	1.515

(D)										
Samples	**20915593**		**T278-50**		**21197826-V1**		**Du151.2**		**19715820_A10_H2**	
3BNC117-LS + 10-1074-LS DP	IC50	IC80	IC50	IC80	IC50	IC80	IC50	IC80	IC50	IC80
	1.525	5.872	1.047	11.952	0.678	2.269	0.004	0.013	0.056	0.204
3BNC117-LS (Control)	>25	>25	>25	>25	>25	>25	>25	>25	>25	>25
10-1074-LS (control)	2.02	5.817	2.174	15.13	0.613	2.188	0.005	0.015	0.074	0.253

3.7. Stability Assessment of Co-Formulated Antibodies

To assess the stability of the co-formulated antibodies at high-concentration in the chosen formulation, we performed a 28 day short stability study that included both a real-time stability study in storage conditions (i.e., at 5 ± 3 °C) and a study of samples in accelerated (25 ± 2 °C/RH 60% ± 5%) and stressed (40 ± 2 °C/RH 75% ± 5%) conditions. RH here refers to Relative Humidity.

For the real-time stability study, we used 2.0 mL co-formulated samples in 3.0 mL Schott glass vials and incubated the samples at 5 ± 3 °C. For evaluation at accelerated and stressed conditions, we used similar sample volumes in 3.0 mL Schott vials and incubated them at 25 ± 2 °C/RH 60% ± 5% (accelerated conditions) and 40 ± 2 °C/RH 75% ± 5% (stressed conditions). Samples were analyzed at T = 0, before study start, and thereafter samples were collected on a weekly basis and analyzed for visual appearance, pH, total protein concentration by UV spectroscopy (280 nm), purity (by determining % monomer and HMW aggregates) by SE-HPLC, charge variants (i.e., relative levels of acidic and basic species) by CEX-HPLC, content of individual antibody by RP-HPLC, protein degradation by SDS-PAGE, sub-visible particles by *FlowCAM®* instrument, viscosity by Viscosizer TD, and (hydrodynamic) particle size by DLS. At real-time storage conditions, the antibodies were stable in the co-formulated milieu for up to 4 weeks across all test parameters (Table 2). In addition, the antibodies were also stable in the accelerated conditions, 25 ± 2 °C/RH 60% ± 5%, for up to 4 weeks across all test parameters (Supplementary Table S4). Furthermore, the co-formulated antibodies were stable up to 4 weeks at stressed conditions, 40 ± 2 °C/RH 75% ± 5%, (Supplementary Table S5).

Table 2. Summary of 28 day stability testing results of co-formulated antibodies, 3BNC117-LS and 10-1074-LS (total 150 mg/mL), evaluated at 0, 1, 2, 3, and 4 weeks, after incubation at storage conditions of 5 ± 3 °C. HMW = High Molecular Weight; d.nm = Diameter in nm; PDI = Polydispersity Index; P/mL = Particles/mL.

Test Attributes		Weeks				
		0	1	2	3	4
pH		5.65	5.6	5.62	5.60	5.59
A280 (mg/mL)		142	137	142	139	149
Viscosity (cP)		10.70	11.08	12.09	11.16	12.89
Osmolality (mOsm/Kg)		345	336	333	336	337
SE-HPLC	HMW (%)	2.98	3.11	3.14	3.58	3.52
	Main Peak (%)	96.90	96.81	96.84	96.41	96.46
CEX-HPLC 3BNC117-LS	Main Peak (%)	48.78	49.08	49.15	49.27	49.71
	Pre-Main Peaks (%)	47.68	46.40	46.04	44.74	44.97
	Post-Main Peaks (%)	3.54	4.52	4.81	5.99	5.32
CEX-HPLC 10-1074-LS	Main Peak (%)	32.68	35.04	35.22	35.99	37.72
	Pre-Main Peaks (%)	61.64	59.70	59.55	59.07	57.45
	Post-Main Peaks (%)	5.68	5.26	5.23	4.94	4.83
RP-HPLC	3BNC117-LS (mg/mL)	68.05	65.88	70.00	68.70	71.48
	10-1074-LS (mg/mL)	76.70	75.75	80.54	78.83	82.09
DLS	Z-Average (d.nm)	10.25	10.44	10.26	10.23	10.30
	PDI	0.18	0.21	0.19	0.18	0.19
FlowCAM	2–10 μm (P/mL)	191	101	253	126	475
	10–25 μm (P/mL)	31	23	46	36	107
	25–50 μm (P/mL)	15	16	8	9	8

In addition to the above analysis, a limited set of samples (T = 2 weeks and T = 4 weeks) from all 3 (real-time, accelerated, and stressed) conditions were tested for potency (functional activity) of the antibodies using pseudovirus neutralization assay. When compared to unincubated co-formulated samples (control) (Table 1A–D), all samples were found to neutralize the pseudoviruses with little to no change in IC50 and 1C80 values and hence found to be stable for up to 4 weeks (Supplementary Table S6A–D). These results indicate not only the utility of the various assays in monitoring antibody stability in the co-formulated sample but also highlight the stability and suitability of the formulation in co-formulating the two antibodies.

This initial stability assessment and identification of appropriate analytical assays support the clinical development of these co-formulated drug products for future clinical studies.

4. Discussion

After the initial evaluation of passive administration of first generation anti-HIV antibodies (4E10+2F5+2G12) [34], the identification of a large number of next-generation anti-HIV bNAbs with greater breadth and potency in the past decade has opened the possibility for antibody-based treatment and/or prevention of HIV-1 infection. Several bNAbs have recently progressed to clinical trials in humans: VRC01 [13,14]/VRC01LS, 10-1074 [15]/10-1074-LS [16], 3BNC117 [17]/3BNC117-LS [18], VRC07-523-LS [19], PGT121 [20,21], and PGDM1400 [21] or their combinations. These early (phase I) clinical studies, with safety, pharmacokinetics, and viral load re-bound or decay as endpoints, have primarily used antibodies formulated for IV infusion. However, to overcome the vast diversity of HIV-1 variants, it is becoming increasingly clear that combinations of (two or more) bNAbs targeting distinct epitopes on the viral envelope (Env) will likely be required [35]. To support the development of bNAb combinations as products for clinical studies, co-formulating two or more antibodies, targeting different Env epitopes, as a single drug product and using a subcutaneous (SC) route for administration, are under consideration in multiple clinical studies. To that end, not only the high-concentration formulation of two or more antibodies in a limited volume will be necessary, but methods to test their individual quality attributes (e.g., purity, charge variants, potency) of the individual antibody in the co-formulated milieu will be required [35].

In this study, two anti-HIV-1 antibodies, 3BNC117-LS and 10-1074-LS, were co-formulated in a 1:1 ratio to achieve a final concentration of 150 mg/mL in 5 mM Histidine, 250 mM Trehalose, 10 mM Methionine, 5 mM Sodium Acetate, 0.05% Polysorbate 20, pH 5.5 buffer. To support the high-concentration formulation and development of the two antibodies for subcutaneous administration, formulation optimizations and analytical test method development and optimizations were performed. Analytical characterization and separation of individual antibodies in the co-formulated sample was challenging due to the high degree of similarity in the physico-chemical properties of the two (3BNC117-LS, and 10-1074-LS) antibodies. Systematic analytical development was carried out, using several methodologies, to obtain separation of the two antibodies. Specifically, chromatographic methods were developed to resolve and assess the quality attributes of the individual antibody in the co-formulated drug product. RP-HPLC and CEX-HPLC methods resulted in baseline separation of the two antibodies (3BNC117-LS and 10-1074-LS) in the co-formulated sample, and the peak profiles compared well to the individually (high-concentration) formulated antibodies. The SE-HPLC method was used to assess the combined high molecular weight species of the two antibodies; the data showed partially separated peaks, corresponding to the two co-formulated antibodies, with no additional % HMW species at this stage. Further evaluation of all HPLC methods for specificity, purity, accuracy, precision, and repeatability confirms that the methods are suitable for future testing of the such co-formulated antibody-based drug product.

In addition to the HPLC methods, an anti-ID ELISA was developed to test identity of the individual antibodies in the co-formulated drug product. In addition, the utilization of a separate and well-defined pseudovirus panel in a virus neutralization assay provided a functional assay platform to not only evaluate the potency/functionality of the individual antibodies but also an approach to test two (or more) antibodies via this functional assay.

In summary, through demonstration of the high-concentration co-formulation of two anti-HIV-1 antibodies and the development of separation-based testing methods, we present several analytical tools to test physico-chemical and functional attributes of co-formulated antibodies, which can contribute to the clinical development of these high-concentration antibodies. Finally, the little to no inter-molecular protein–protein interaction between the antibodies, even at ≥150 mg/mL, and their stability profile ensure the possibility of the development of such high-concentration antibodies as products for HIV prevention and/or treatment.

Supplementary Materials: The following are available online at http://www.mdpi.com/2073-4468/9/3/36/s1: Table S1: Precision and accuracy of the reverse phase chromatography method during analysis of co-formulated antibodies. Intra- and Inter- assay precision of (A) total peak area (10-1074-LS and 3BNC117-LS peaks), (B) 10-1074-LS-specific peak area and (C) 3BNC117-LS-specific peak area from six replicate runs of co-formulated samples analyzed by reverse phase chromatography. (D) Accuracy determination from percent recovery of a range of co-formulated samples analyzed on three different days (day 1, 2 and 3); Table S2: Precision and accuracy of the cation-exchange chromatography method during analysis of co-formulated antibodies. Intra- and Inter- assay precision of (A) total peak area (10-1074-LS and 3BNC117-LS peaks), (B) 10-1074-LS-specific peak area and (C) 3BNC117-LS-specific peak area from six replicate runs of co-formulated samples analyzed by cation-exchange chromatography. (D) Accuracy determination from percent recovery of a range of co-formulated samples analyzed on three different days (day 1, 2 and 3) for 6 different concentrations (50, 80, 100, 120, 200 and 300 g); Table S3: Precision and accuracy of the size-exclusion chromatography method during analysis of co-formulated antibodies. Intra- and Inter- assay precision of (A) main peak area (10-1074-LS and 3BNC117-LS monomer), (B) High Molecular Weight (HMW) peak area and (C) Low Molecular Weight (LMW) peak area from six replicate runs of co-formulated samples analyzed by size-exclusion chromatography. (D) Accuracy determination from percent recovery of a range of co-formulated samples analyzed on three different days (day 1, 2 and 3) for 6 different concentrations (50, 80, 100, 120, 200 and 300 g); Table S4: Summary of 28-day stability testing results of co-formulated antibodies, 3BNC117-LS and 10-1074-LS (each at 75 mg/mL), evaluated at 0, 1, 2, 3, and 4 weeks, after incubation at accelerated conditions of 25 ± 2 °C/RH 60% ± 5%. HMW = High Molecular Weight; PDI = Polydispersity Index; P/mL = Particles/mL; Table S5: Summary of 28-day stability testing results of co-formulated antibodies, 3BNC117-LS and 10-1074-LS, evaluated at 0, 1, 2, 3, and 4 weeks, after incubation at stressed conditions of 40 ± 2 °C/75 ± 5% RH. HMW = High Molecular Weight; PDI = Polydispersity Index; P/mL = Particles/mL; Table S6: Testing of functional activity of 3BNC117-LS and 10-1074-LS in the co-formulated samples from the 28-days stability study using pseudovirus neutralization assay. 2 weeks samples (T = 2 weeks) and 4 weeks samples (T = 4 weeks) were selected from all 3 conditions and analyzed against 2 panels of pseudoviruses in TZM-bl cells. One panel (A and B) is used to test functional activity of 3BNC117 [3BNC117 sensitive/10-1074 resistant viruses (n = 10)] and the other panel (C and D) is used to test functional activity of 10-1074 [3BNC117 resistant/10-1074 sensitive viruses (n = 10)]. Both IC50 and IC80 values (in g/mL) are reported.

Author Contributions: A.K.D. and I.J. designed the experiments and strategy. B.M., K.T.M., and K.N. performed the experiments. V.K.S., S.A., K.N., M.C., J.H., M.C.N., M.S.S., I.J., and A.K.D. reviewed the data. V.K.S., B.M., and A.K.D. wrote the manuscript. All authors have read and agreed to the published version of the manuscript.

Funding: This work was funded by Bill and Melinda Gates Foundation (BMGF) Collaboration for AIDS Vaccine Development (CAVD), grant number OPP1147661 and OPP1153692 to International AIDS Vaccine Initiative (IAVI) and grant number OPP1146996 to the Comprehensive Antibody Vaccine Immune Monitoring Consortium.

Acknowledgments: We would like to thank Kirill Yakovlevsky, Charles McneTablemar, and Amina Soukrati (at CuriRx, Inc., 205 Lowell Street, Wilmington, MA 01887, USA) for excellent technical assistance. We are grateful to Pervin Anklesaria and Susan Barnett (BMGF) for their input and support in this project.

Conflicts of Interest: The authors do not have any conflicts of interest to declare.

References

1. Singh, S.; Kumar, N.K.; Dwiwedi, P.; Charan, J.; Kaur, R.; Sidhu, P.; Chugh, V.K. Monoclonal Antibodies: A Review. *Curr. Clin. Pharmacol.* **2018**, *13*, 85–99. [CrossRef] [PubMed]
2. Marston, H.D.; Paules, C.I.; Fauci, A.S. Monoclonal Antibodies for Emerging Infectious Diseases—Borrowing from History. *N. Engl. J. Med.* **2018**, *378*, 1469–1472. [CrossRef] [PubMed]
3. Walker, L.M.; Burton, D.R. Passive immunotherapy of viral infections: 'super-antibodies' enter the fray. *Nat. Rev. Immunol.* **2018**, *18*, 297–308. [CrossRef] [PubMed]
4. Keeffe, J.R.; Van Rompay, K.K.A.; Olsen, P.C.; Wang, Q.; Gazumyan, A.; Azzopardi, S.A.; Schaefer-Babajew, D.; Lee, Y.E.; Stuart, J.B.; Singapuri, A.; et al. A Combination of Two Human Monoclonal Antibodies Prevents Zika Virus Escape Mutations in Non-human Primates. *Cell Rep.* **2018**, *25*, 1385–1394. [CrossRef] [PubMed]
5. Geevarghese, B.; Simoes, E.A. Antibodies for prevention and treatment of respiratory syncytial virus infections in children. *Antivir. Ther.* **2012**, *17*, 201–211. [CrossRef] [PubMed]
6. Bittner, B.; Richter, W.; Schmidt, J. Subcutaneous Administration of Biotherapeutics: An Overview of Current Challenges and Opportunities. *BioDrugs* **2018**, *32*, 425–440. [CrossRef]
7. Stoner, K.L.; Harder, H.; Fallowfield, L.J.; Jenkins, V.A. Intravenous versus subcutaneous drug administration. Which do patients prefer? A systematic review. *Patient* **2015**, *8*, 145–153. [CrossRef]
8. Stephenson, K.E.; Barouch, D.H. Broadly Neutralizing Antibodies for HIV Eradication. *Curr. HIV/AIDS Rep.* **2016**, *13*, 31–37. [CrossRef]
9. Pegu, A.; Hessell, A.J.; Mascola, J.R.; Haigwood, N.L. Use of broadly neutralizing antibodies for HIV-1 prevention. *Immunol. Rev.* **2017**, *275*, 296–312. [CrossRef]
10. Kwong, P.D.; Mascola, J.R. HIV-1 Vaccines Based on Antibody Identification, B Cell Ontogeny, and Epitope Structure. *Immunity* **2018**, *48*, 855–871. [CrossRef]
11. McMichael, A.J.; Haynes, B.F. Lessons learned from HIV-1 vaccine trials: New priorities and directions. *Nat. Immunol.* **2012**, *13*, 423–427. [CrossRef] [PubMed]
12. Haynes, B.F.; McElrath, M.J. Progress in HIV-1 vaccine development. *Curr. Opin. HIV AIDS* **2013**, *8*, 326–332. [CrossRef] [PubMed]
13. Ledgerwood, J.E.; Coates, E.E.; Yamshchikov, G.; Saunders, J.G.; Holman, L.; Enama, M.E.; DeZure, A.; Lynch, R.M.; Gordon, I.; Plummer, S.; et al. Safety, pharmacokinetics and neutralization of the broadly neutralizing HIV-1 human monoclonal antibody VRC01 in healthy adults. *Clin. Exp. Immunol.* **2015**, *182*, 289–301. [CrossRef] [PubMed]
14. Lynch, R.M.; Boritz, E.; Coates, E.E.; DeZure, A.; Madden, P.; Costner, P.; Enama, M.E.; Plummer, S.; Holman, L.; Hendel, C.S.; et al. Virologic effects of broadly neutralizing antibody VRC01 administration during chronic HIV-1 infection. *Sci. Transl. Med.* **2015**, *7*, 319ra206. [CrossRef]

15. Caskey, M.; Schoofs, T.; Gruell, H.; Settler, A.; Karagounis, T.; Kreider, E.F.; Murrell, B.; Pfeifer, N.; Nogueira, L.; Oliveira, T.Y.; et al. Antibody 10-1074 suppresses viremia in HIV-1-infected individuals. *Nat. Med.* **2017**, *23*, 185–191. [CrossRef]
16. First-in-human Study of 10-1074-LS Alone and in Combination with 3BNC117-LS. Available online: https://clinicaltrials.gov/ct2/show/NCT03554408 (accessed on 28 July 2020).
17. Scheid, J.F.; Horwitz, J.A.; Bar-On, Y.; Kreider, E.F.; Lu, C.L.; Lorenzi, J.C.; Feldmann, A.; Braunschweig, M.; Nogueira, L.; Oliveira, T.; et al. HIV-1 antibody 3BNC117 suppresses viral rebound in humans during treatment interruption. *Nature* **2016**, *535*, 556–560. [CrossRef]
18. 3BNC117-LS First-in-Human Phase 1 Study. Available online: https://clinicaltrials.gov/ct2/show/NCT03254277 (accessed on 28 July 2020).
19. Evaluating the Safety and Pharmacokinetics of VRC01, VRC01LS, and VRC07-523LS, Potent Anti-HIV Neutralizing Monoclonal Antibodies, in HIV-1-Exposed Infants. Available online: https://clinicaltrials.gov/ct2/show/NCT02256631 (accessed on 28 July 2020).
20. Safety, PK and Antiviral Activity of PGT121 Monoclonal Antibody in HIV-uninfected and HIV-infected Adults. Available online: https://clinicaltrials.gov/ct2/show/NCT02960581 (accessed on 28 July 2020).
21. A Clinical Trial of PGDM1400 and PGT121 and VRC07-523LS Monoclonal Antibodies in HIV-infected and HIV-uninfected Adults. Available online: https://clinicaltrials.gov/ct2/show/NCT03205917 (accessed on 28 July 2020).
22. Bar-On, Y.; Gruell, H.; Schoofs, T.; Pai, J.A.; Nogueira, L.; Butler, A.L.; Millard, K.; Lehmann, C.; Suarez, I.; Oliveira, T.Y.; et al. Safety and antiviral activity of combination HIV-1 broadly neutralizing antibodies in viremic individuals. *Nat. Med.* **2018**, *24*, 1701–1707. [CrossRef]
23. Mendoza, P.; Gruell, H.; Nogueira, L.; Pai, J.A.; Butler, A.L.; Millard, K.; Lehmann, C.; Suarez, I.; Oliveira, T.Y.; Lorenzi, J.C.C.; et al. Combination therapy with anti-HIV-1 antibodies maintains viral suppression. *Nature* **2018**, *561*, 479–484. [CrossRef]
24. Halper-Stromberg, A.; Nussenzweig, M.C. Towards HIV-1 remission: Potential roles for broadly neutralizing antibodies. *J. Clin. Investig.* **2016**, *126*, 415–423. [CrossRef]
25. Mueller, C.; Altenburger, U.; Mohl, S. Challenges for the pharmaceutical technical development of protein coformulations. *J. Pharm. Pharmacol.* **2018**, *70*, 666–674. [CrossRef]
26. Cao, M.; De Mel, N.; Shannon, A.; Prophet, M.; Wang, C.; Xu, W.; Niu, B.; Kim, J.; Albarghouthi, M.; Liu, D.; et al. Charge variants characterization and release assay development for co-formulated antibodies as a combination therapy. *MAbs* **2019**, *11*, 489–499. [CrossRef] [PubMed]
27. Patel, A.; Gupta, V.; Hickey, J.; Nightlinger, N.S.; Rogers, R.S.; Siska, C.; Joshi, S.B.; Seaman, M.S.; Volkin, D.B.; Kerwin, B.A. Coformulation of Broadly Neutralizing Antibodies 3BNC117 and PGT121: Analytical Challenges During Preformulation Characterization and Storage Stability Studies. *J. Pharm. Sci.* **2018**, *107*, 3032–3046. [CrossRef] [PubMed]
28. Ko, S.Y.; Pegu, A.; Rudicell, R.S.; Yang, Z.Y.; Joyce, M.G.; Chen, X.; Wang, K.; Bao, S.; Kraemer, T.D.; Rath, T.; et al. Enhanced neonatal Fc receptor function improves protection against primate SHIV infection. *Nature* **2014**, *514*, 642–645. [CrossRef] [PubMed]
29. Zalevsky, J.; Chamberlain, A.K.; Horton, H.M.; Karki, S.; Leung, I.W.; Sproule, T.J.; Lazar, G.A.; Roopenian, D.C.; Desjarlais, J.R. Enhanced antibody half-life improves in vivo activity. *Nat. Biotechnol.* **2010**, *28*, 157–159. [CrossRef]
30. Montefiori, D.C. Measuring HIV neutralization in a luciferase reporter gene assay. *Methods Mol. Biol.* **2009**, *485*, 395–405.
31. Sarzotti-Kelsoe, M.; Bailer, R.T.; Turk, E.; Lin, C.L.; Bilska, M.; Greene, K.M.; Gao, H.; Todd, C.A.; Ozaki, D.A.; Seaman, M.S.; et al. Optimization and validation of the TZM-bl assay for standardized assessments of neutralizing antibodies against HIV-1. *J. Immunol. Methods* **2014**, *409*, 131–146. [CrossRef]
32. Fekete, S.; Beck, A.; Fekete, J.; Guillarme, D. Method development for the separation of monoclonal antibody charge variants in cation exchange chromatography, Part II: pH gradient approach. *J. Pharm. Biomed. Anal.* **2015**, *102*, 282–289. [CrossRef]
33. Fekete, S.; Beck, A.; Fekete, J.; Guillarme, D. Method development for the separation of monoclonal antibody charge variants in cation exchange chromatography, Part I: Salt gradient approach. *J. Pharm. Biomed. Anal.* **2015**, *102*, 33–44. [CrossRef]

34. Armbruster, C.; Stiegler, G.M.; Vcelar, B.A.; Jager, W.; Koller, U.; Jilch, R.; Ammann, C.G.; Pruenster, M.; Stoiber, H.; Katinger, H.W. Passive immunization with the anti-HIV-1 human monoclonal antibody (hMAb) 4E10 and the hMAb combination 4E10/2F5/2G12. *J. Antimicrob. Chemother.* **2004**, *54*, 915–920. [CrossRef]
35. Wagh, K.; Bhattacharya, T.; Williamson, C.; Robles, A.; Bayne, M.; Garrity, J.; Rist, M.; Rademeyer, C.; Yoon, H.; Lapedes, A.; et al. Optimal Combinations of Broadly Neutralizing Antibodies for Prevention and Treatment of HIV-1 Clade C Infection. *PLoS Pathog.* **2016**, *12*, e1005520. [CrossRef]

Review

Antibodies Inhibiting the Type III Secretion System of Gram-Negative Pathogenic Bacteria

Julia A. Hotinger and Aaron E. May *

Department of Medicinal Chemistry, School of Pharmacy, Virginia Commonwealth University, Richmond, VA 23219, USA; hotingerja@vcu.edu

* Correspondence: aemay@vcu.edu; Tel.: +1-804-828-7134

Received: 10 June 2020; Accepted: 22 July 2020; Published: 27 July 2020

Abstract: Pathogenic bacteria are a global health threat, with over 2 million infections caused by Gram-negative bacteria every year in the United States. This problem is exacerbated by the increase in resistance to common antibiotics that are routinely used to treat these infections, creating an urgent need for innovative ways to treat and prevent virulence caused by these pathogens. Many Gram-negative pathogenic bacteria use a type III secretion system (T3SS) to inject toxins and other effector proteins directly into host cells. The T3SS has become a popular anti-virulence target because it is required for pathogenesis and knockouts have attenuated virulence. It is also not required for survival, which should result in less selective pressure for resistance formation against T3SS inhibitors. In this review, we will highlight selected examples of direct antibody immunizations and the use of antibodies in immunotherapy treatments that target the bacterial T3SS. These examples include antibodies targeting the T3SS of *Pseudomonas aeruginosa*, *Yersinia pestis*, *Escherichia coli*, *Salmonella enterica*, *Shigella* spp., and *Chlamydia trachomatis*.

Keywords: type III secretion system; antibodies; prophylaxis; antibacterials; antibiotics

1. Introduction

The type III secretion system (T3SS) is a multimeric protein complex used by many pathogenic Gram-negative bacteria to cause and maintain an infection [1]. Pathogens that use a T3SS include *Chlamydia trachomatis*, *Escherichia coli*, *Pseudomonas aeruginosa*, *Salmonella enterica*, *Shigella* spp., *Vibrio cholerae*, and *Yersinia pestis* [2]. The T3SS functions as a molecular syringe, sometimes called an injectisome that bacteria use to translocate effector proteins directly into a host cell (Figure 1) [3]. The T3SS is comprised of three major components. First, a basal body that anchors the structure to the bacterial membrane containing an ATPase at the base that powers the secretion of proteins. Next, the needle itself acts as a tunnel that spans the extracellular space between the pathogen and host cell. Finally, there is a translocon that forms a pore in the host cell membrane [4]. Due to the small diameter of the needle, the effector proteins must be unfolded to be translocated and then are re-folded after entering the host cell [5]. These effector proteins are responsible for modifying the host cell functions in ways that are beneficial to the pathogen. This includes mechanisms such as reprogramming host machinery to allow for colonization through interference with actin and tubulin, gene expression, or cell cycle progression (*Salmonella* spp., *Shigella* spp.) [6,7]. Some pathogens even interfere with or induce programmed cell death (*Yersinia* spp., *Pseudomonas* spp.) [8,9].

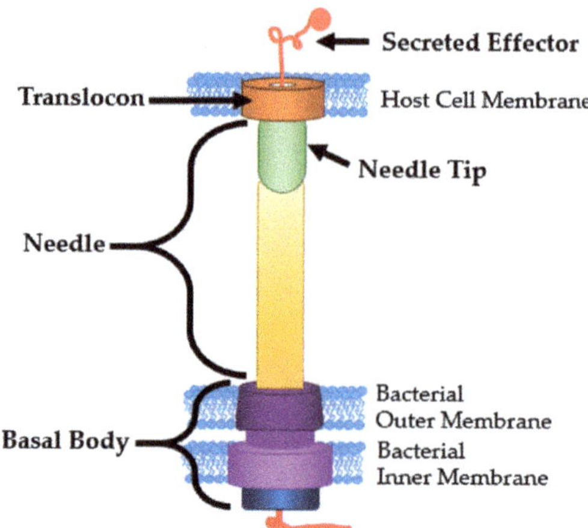

Figure 1. T3SS Structure and Common Targets. Modified from [3].

The T3SS is becoming an important anti-virulence target for many reasons. The T3SS is specific to Gram-negative pathogens, meaning any interventions targeting it should not affect commensal bacteria [10]. Bacteria containing a nonfunctional T3SS also have attenuated virulence but are still capable of growth [11–15]. This lends to the theory that inhibiting the T3SS will reduce the selective pressure on the bacterial pathogen to form resistance, leading to slower formation of resistance to T3SS inhibitors [16]. Small molecule inhibitors of the T3SS have been shown to increase survival rates after infection with otherwise lethal doses of bacterial pathogens [10,17].

Mammalian immune systems produce antibodies (Ab) against T3SS proteins when natural infection occurs [18–22]. Due to the high prevalence of infection caused by bacteria utilizing the T3SS, the majority of humans have antibodies to the T3SS of some pathogens already in their system [22]. Durand et al. tested human colostrum samples for Abs against T3SS proteins for *Salmonella* spp., *Shigella* spp., and *E. coli* including the needle tip, translocon, and secreted effectors. They found that every sample collected contained Abs to at least one of the aforementioned proteins and 10% of the samples contained Abs to all 11 proteins tested [22]. When pregnant cattle were vaccinated against *E. coli* with two recombinant T3SS-related proteins, EspB, and γ-intimin, the Abs produced against these antigens was passed to their calves through breast milk [23]. Rabinovitz et al. showed that calves with vaccinated mothers showed markedly higher survival rates after a challenge of enterohemorrhagic *E. coli* (EHEC) than those with sham-vaccinated mothers [24].

Antibody recognition can lead to rapid and robust responses by the immune system, removing the pathogen before any symptoms can be felt or seen in the host. When this is the case, we consider the host to be immune to the pathogen [25]. The presence of anti-T3SS Abs is enough to identify that an individual has come into contact with the pathogen with the T3SS protein in question, but not necessarily that they have immunity. This is because not all Abs have the same immunoprotective properties [26]. Notwithstanding this fact, the presence of the Abs targeting the T3SS and its effectors across multiple bacterial species implies a significant therapeutic potential. The most T3SS structural components are not expressed by non-pathogenic bacteria, allowing for the potential for enhanced specificity. In this review, we will cover selected examples of promising and effective antibody-based treatments and prophylactics that target the T3SS of pathogenic bacteria.

2. Antibody Structure and Function

The majority of antibodies are "Y" shaped immunoglobulin (Ig) proteins that are used by the immune system to recognize antigens. They contain a variable domain on the tips on the Y and bind to antigens. A non-variable or constant domain on the stem of the Y binds to cellular receptors

(Figure 2A) [25]. There are five main isotypes of Abs found in humans: IgA, IgD, IgE, IgM and IgG. IgA and IgG are the most commonly used in therapeutics [25,27,28]. IgA are found in the mucosal membranes and help to prevent the colonization of mucosal pathogens. They are commonly found as dimers that take the shape of two Y's bound together at the stem [27]. IgG are considered memory Abs and provide the main Ab-based immunity against pathogens and comprise approximately 80% of total pooled Abs within humans [28]. Some antibodies, such as IgD, are membrane-bound and involved in cellular signaling. Nearly all antibodies are glycosylated to assist in specificity and binding. There are two main types of Abs in the context of antigen binding. Monoclonal antibodies (mAbs) are identical in their sequence and specificity, while polyclonal antibodies (pAbs) are not identical in sequence [25]. MAbs are more often used as therapeutics and vaccines due to their higher specificity and homogeneity [29,30].

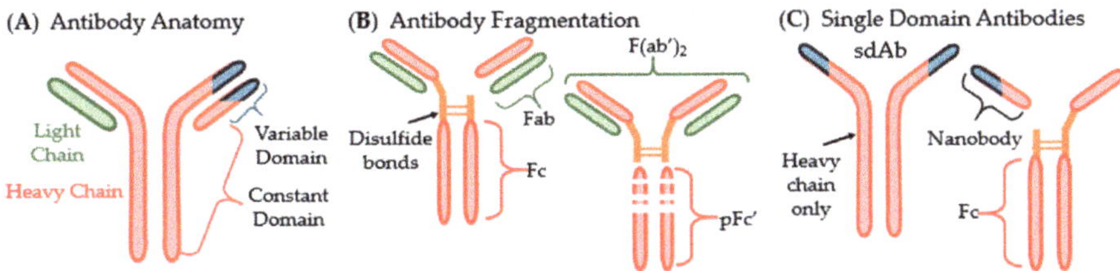

Figure 2. Structure of antibody and common fragmentation types: (**A**) generic anatomy of an antibody; (**B**) visualization of fragmented antibodies; (**C**) single domain antibodies (sdAbs) and their fragmentation.

Antigen binding fragments (Fabs or Fvs) can often be used in place of whole antibodies. These fragments are one light chain and a section of a whole Ab that contains the variable domain. Another type of fragment called $F(ab')_2$, essentially two Fabs linked together, can also be isolated from Ab solutions (Figure 2B). Another Ab type that has become important in pharmaceutical development is small- or single-domain antibodies (sdAb) which were discovered in the family *Camelidae* (Figure 2C) [31,32]. These sdAb consist of the heavy chain homodimers that lack the light chains entirely. SdAbs are often cleaved at their disulfide bonds to separate the variable domain from the Fc region. The variable domain fragments are also called single variable domains (VHH, scFv) because their antigen-binding site is a singular variable domain of a heavy chain IgG. VHHs are also often called nanobodies due to their small size (Nanobody™ is a trademark of Ablynx N.V., Ghent, Belgium) [32].

There are multiple mechanisms by which an antibody can act to destroy or inactivate infectious agents. These include: (1) Complement-dependent bacteriolysis; (2) Opsonization or phagocytosis; (3) Antibody-dependent cell-mediated cytotoxicity (ADCC); (4) Agglutination; (5) Neutralization; and (6) Secretion blockade (Figure 3) [25,33,34]. Mechanisms 1–3 and 5 were the original biological effects that Abs were thought to perform. Agglutination (4) is not typically considered one of the mechanisms of Abs against bacterial pathogens because it leads to mechanism 2 or 3 but is included here for clarity. Secretion blockades (6) were only recently discovered in the context of T3SS inhibition research.

Complement activation occurs when Abs bind to an antigen of the bacteria or virus. This attracts the first component of the complement cascade and subsequently the classical complement system [33]. This activation results in pathogen death. The entire process is called complement-dependent cytotoxicity (CDC). CDC is divided into two distinct pathways for pathogen elimination; the Ab attracts and begins the formation of a membrane attack complex which then assists in bacteriolysis (1) or the Ab marks the bacteria for opsonization by neutrophils, macrophages, or other phagocytes (2) [33]. Opsonization is considered an indirect inactivation or inhibition of pathogenesis by Abs. This is because the Ab itself does not cause the halt of pathogenesis. Along with eventual bacterial death, complement activation also attracts inflammatory cells to the site [25].

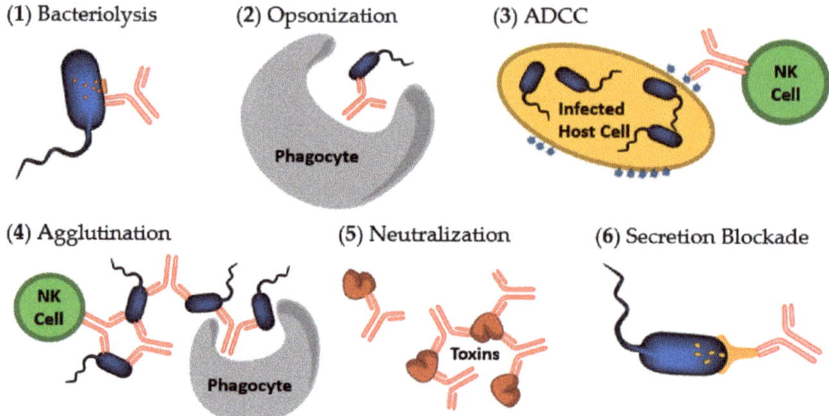

Figure 3. Mechanisms initiated by antibodies to destroy bacteria or toxins. (**1**) Bacteriolysis occurring after complement activation; (**2**) Opsonization by a macrophage or neutrophil after Fc sequence recognition; (**3**) Antibody-dependent cell-mediated cytotoxicity (ADCC) of an infected host cell; (**4**) Agglutination; (**5**) Neutralization of a bacterial secreted toxin; (**6**) Secretion blockade preventing T3SS proteins from being secreted. Image modified from [34].

ADCC is initiated by Abs that mark infected host cells for digestion or lysis (3). For example, after a pathogen invades a host cell, the host cell may break up some of the pathogen's proteins and display them on the host membrane. Abs can bind to the displayed pathogen protein fragments and be recognized by natural killer (NK) cells. The NK cells then induce apoptosis of the infected host cell [35]. For clarity within this review, Abs that attach to antigens presented on the host cell will be considered marked for ADCC. When the bacterial cell is attached to the host cell then either ADCC or opsonization can be considered.

Agglutination occurs when the Ab binds to multiple foreign cells, clumping them together into large attractive targets for phagocytes (4). This eventually leads to opsonization (2). Agglutination also activates natural killer cells and initiates ADCC (3) [33,35]. Agglutination helps to prevent cell division in bacterial pathogens by physically lumping the cells together [25].

Neutralization is the process in which an Ab binds to an antigen, typically a toxin, causing physical or chemical inactivation of that antigen (5) [33]. Precipitation is another specific way in which antigens can be neutralized. Abs may bind to multiple soluble antigens to create larger, insoluble clumps that precipitate out of solution, once again making them attractive targets for phagocytes [33].

The most recently discovered mechanism for deactivating pathogens are secretion blockades. A secretion blockade occurs when the Ab binds to a secretion system and physically blocks the secretion of protein (6). This helps to prevent the bacteria from binding to host cells and infecting them [34]. This mechanism is initiated by Abs targeting the translocon or needle tip proteins of the T3SS. When Abs latch onto these proteins it can create a physical barrier, preventing the needle tip from attaching to the translocon correctly or the translocon from integrating into host cell membranes. One example of this phenomenon is seen by specific anti-LcrV Ab blocking the apoptotic action of LcrV, the *Yersinia* spp. needle tip, against human T-cells [36].

3. Antibodies as Pharmaceuticals

Edward Jenner, the father of the modern vaccine, used the blood serum of milkmaids who were immune to smallpox due to their exposure to cowpox to successfully vaccinate a child against smallpox [37]. This strategy was inspired by the way infants receive protection from maternal antibodies contained in their mother's milk and was considered to be a passive immunization [23,38]. Since Jenner's time, passive immunization has come to mean the direct administration of purified antibodies or antibody serum rather than human blood containing antibodies [30]. In contrast, active immunization is done with an antigen, such as a toxoid, or with whole-cell vaccines. The passive distinction comes from the lack of immune response required by the host to confer immunity.

Any vaccine can become ineffective over time due to mutations of the bacteria or virus, but when the protein's sequence is highly conserved the chance of mutation is decreased [38]. Many components of the T3SS, such as the translocon, needle tip, needle subunits, and ATPases, are highly conserved between strains of a species of bacteria and often even between species of a particular genus [2]. This conservation allows for the immune response elicited by the injected antigen to have a high likelihood of recognition amongst different species or serovars of bacteria within the same genus. Vaccines targeting the T3SS also have shown promise as some subunit vaccines of the *Yersinia* needle tip protein have gone into clinical trials [39].

Nonspecific polyclonal human IgG pooled from 10,000 s of donors is known as intravenous immunoglobulin (IVIG) and has a wide variety of clinical uses that highlight the importance of using IgG as a therapeutic. IVIG contains antibodies with the ability to bind to a wide variety of antigens because it is pooled from so many donors [40,41]. IVIG is often used in immunocompromised patients as a prophylactic to prevent infection [42,43], or in those struggling with active infections [44,45]. If a grouping of non-specific antibodies can help to prevent and treat disease, then antibodies designed specifically to act upon pathogens should be able to do the same at a higher specificity without some of the common side effects and complications of IVIG, such as fever, migraines, anxiety, nausea, and vomiting [46,47]. Other benefits to designed antibody therapeutics are the consistency in their preparation, homogeneity of contents, and ease of engineering [48].

Antibodies have recently come into prominence as therapeutics. To date, there have been approximately 85 mAbs approved by the FDA for use as immunotherapy and 80 that have been approved by countries within the European Union [49]. Abs used as therapeutics are often administered via intravenous (IV) injections, intramuscularly, or parenterally [28,47]. The route of administration can be very important for the effectiveness of therapeutic Abs. For example, Sécher et al. showed the administration of anti-PcrV pAbs is more effective at treating *Pseudomonas aeruginosa* infections when done via airways than through parenteral injections [50]. This is likely due to the localization of the therapeutic at the area of infection as *P. aeruginosa* infects lung epithelial tissue. In the case of a gastrointestinal pathogen, such as *E. coli* or *Shigella* spp., an oral route of administration may be more effective than IV injections [51]. Hill et al. showed that local administrations of anti-LcrV and F1 Abs could be used as a *Yersinia* infection treatment or prophylactic, while injections of the same Abs could only act preventatively when administered multiple weeks in advance [52]. Their research suggests that this method of localization to the lung could be effective as a fast-acting post-exposure treatment for pneumonic plague.

Breastfeeding has been shown to reduce the risk of infant diarrheal disease, of which *E. coli* is a main culprit, from 76% to 26% [53]. Loureiro et al. showed that passive immunization of infants with anti-T3SS Abs via breastfeeding protects them against infection with two strains of EPEC. Abs targeting three separate T3SS-related proteins were discovered in infants in areas where EPEC-caused diarrhea is endemic. These Abs were isolated from blood samples and shown to decrease host cell binding of EPEC. They also acted as potent opsonins for killing EPEC [18]. Both of these are evidence for the protective properties of anti-T3SS Abs against Gram-negative pathogenic bacteria.

No studies on Abs protecting against a bacterial challenge have been performed in human infants. There have, however, been studies in baboons. Kapil et al.'s study of maternal vaccination against *Bordetella pertussis*, the causative agent of whooping cough, showed that Abs transferred via breastfeeding were sufficient to protect against a *B. pertussis* infection. Infant baboons born to vaccinated mothers did become highly colonized with the pathogen but did not exhibit signs of disease and cleared the infectious bacteria approximately three weeks after bacterial challenge. The infants born to non-vaccinated mothers, on the other hand, exhibited severe disease symptoms and all but one were euthanized due to the severity of symptoms [54].

3.1. Challenges of Anti-T3SS Antibody Therapies

Compared to small molecules, Abs have much higher specificity and affinity to their targets and can easily be identified as drug candidates. They also have longer half-lives due to lower CYP450 metabolism and high serum stability [29]. Along with these benefits, Abs are twice as likely to be approved and moved to market once entering in-human trials than small molecules [55]. All of these traits make Abs desirable candidates for drug development. Unfortunately, there are some downsides to Ab therapeutics when compared to small molecules. Abs typically cannot penetrate cell membranes, which means that intracellular targets are unavailable to them [56]. This issue has inspired techniques to engineer cell-penetrating Abs and antibody fragments, but these increase the cost of production [57]. Abs also have a higher cost of production than small molecules [58].

When discussing antibody therapies the risk of host rejection and severe side effects must also be considered. A higher dosage leads to a higher risk of adverse side effects or effects on plasma viscosity. This is most commonly an issue with IVIG where replacement treatments are approximately 200–400 mg/kg given every two to three weeks continuously and acute treatments can reach 2000 mg/kg monthly. The large doses in IVIG treatment are less desirable when compared to the treatment of acute infections with mAbs that range closer to 5–50 mg/kg [59]. This can be contrasted, however, with the risk of anti-antibody formation.

Prolonged use of biologics, particularly mAbs, can cause the development of anti-drug antibodies (ADAs). IVIG is at low risk for ADA neutralization because the solution of Abs is from separate donors and less likely to contain high concentrations of any particular antibody. In comparison, mAbs are a singular Ab, meaning only one type of ADA is required for neutralization. Once ADA are present in the patient the treatment is often no longer administered due to changes in pharmacokinetics (PK) and the risk of allergic reactions. ADAs can function by a variety of different mechanisms. Some ADAs, called binding ADAs, increase clearance by complex formation while others, neutralizing ADAs, increase clearance and neutralize by binding to the epitope associated with the therapy [59]. Once a biologic therapy reaches preclinical status, the effect of ADAs on immunogenicity and PK must be considered and determined. The risk of an allergic reaction due to ADA formation must also be documented. The production of ADAs against an anti-T3SS antibody would not necessarily be prohibitory to its success. The risk of ADAs against mAbs currently on the market range from 0% to 89%, although the majority are under 10% [59].

Clinical trials of antibodies used to treat Gram-negative infections, including sepsis, have had limited success in the past. IVIG has been approved to treat sepsis, but there is a high degree of heterogeneity in the results of treatment, leading to unclear guidelines for dosing and preparation [48]. MAbs have faced even larger difficulties. A clinical trial on the effects of anti-lipopolysaccharide (LPS) IgG revealed no therapeutic benefit in any clinical parameter measured from either the administration of anti-LPS mAbs or endogenously produced antibodies [60]. It is unclear whether therapeutic antibodies targeting the T3SS will encounter the same difficulties. A clinical trial on an anti-PcrV $F(ab')_2$ was tested for its efficacy at preventing and treating sepsis due to *P. aeruginosa*. The treatment did significantly prevent onset of sepsis, but protection was not considered complete. When administered after the onset of sepsis there was a slowing of disease progression, resulting in a decreased rate of septic shock. Presence of the pathogen, however, was not significantly decreased by the administration of anti-PcrV $F(ab')_2$ [61]. This suggests that anti-T3SS antibody treatments alone may not be enough to treat or prevent disease.

The research presented in this review is focused on the use of anti-T3SS Abs as individual therapies, but combination therapies are more likely in practice [62,63]. Combination therapy may reduce drug resistance emergence by allowing for reduced dosages and treatment duration of antibiotics [64]. Evidence of this synergistic approach can be seen in Secher et al.'s study of an anti-*P. aeruginosa* mAb with meropenem, a broad-spectrum antibiotic. They found treatment with this combination led to an additive effect. When the combination was given to patients with meropenem-resistant infections the mAb efficacy was comparable to treatments with the mAb alone against meropenem-sensitive infections [65]. Le at al. demonstrated that MEDI3902, an anti-T3SS mAb, showed enhanced activity when treating *P. aeruginosa* infections when administered in combination with a subtherapeutic dose of meropenem [66]. These studies, along with others discussed in this review, are evidence of practical applications of anti-T3SS Ab therapeutics when used in combination with traditional antibiotic approaches.

3.2. Strategies to Enhance Antibody Production

The costs associated with antibody production and isolation can be a limiting factor in their use. Strategies to enhance and reduce the cost of antibody therapeutics is an ongoing research area. Since the mid-1970s, hybridomas have been the main technology used in mAb production. The most common cells used for recombinant mAb production are Chinese hamster ovary (CHO) cells [67,68]. Identification and engineering of high-Ab production cell lines have long been a challenge. CHO cells are known to produce antibodies at approximately 1 g/L after optimization [68,69]. Factors that play a role in the success and efficiency of cell lines include the time until desired cell density is reached, the duration of production time allowing for antibody harvesting, and the overall titer of antibody produced [67]. Itoh et al. found that suppression of apoptosis-associated genes allows for longer culture times and therefore higher Ab titers [70]. The overproduction of proteins involved in protein folding, such as CHOP, have also shown improvement in the viability of antibodies produced [71]. Sittner et al. developed fluorescence-activated cell sorting (FACS) to produce Abs more effectively against LcrV, a T3SS needle tip protein. This technique is intended to be used in conjunction with hybridoma mAb production. FACS increased yields of mAbs by 773%, from 22 to 170 positive clones per spleen [72].

MAbs are routinely produced by hybridomas; replacing them with bacteria is a strategy to reduce the cost of production [73,74]. MAbs produced in bacteria face challenges of protein aggregation, inefficient folding, and low yields [75]. Zhou et al. produced a full-length mAb in *E. coli* by fusing the signal peptide of disulfide oxidoreductase to the N-terminus of the heavy chain of the mAb to assist in secretion into and accumulation in the periplasm. This tag helped to reduce the bottleneck in production caused by inefficient heavy chain secretion [73]. Plants have also been used to produce Abs cost-effectively. Saberianfar et al. isolated sdAbs that target a T3SS effector from tobacco (*Nicotiana benthamiana*) leaves at a level of 1% to 3% of total soluble protein [74].

4. T3SS Components Targeted by Antibodies

The two most common targets of antibodies against the T3SS are the needle tip and translocon proteins. The binding sites of Abs to either of these proteins are easily accessible to the antibodies due to their extracellular location. Binding marks the bacteria for opsonization or creates a secretion blockade to prevent effectors from entering the host cell. Part of the basal body is available extracellularly and has been used as an Ab target. Secreted effector proteins are also common targets of Ab therapies because they are often toxins and humans naturally produce Abs against them. There has been limited experimentation targeting regulatory proteins to prevent expression of the T3SS.

4.1. Needle Tip

The needle tip protein of many T3SSs, also called the V antigen, causes mammalian hosts to produce specific IgG. Kinoshita et al. found relatively high and comparable antibody titers in human sera against V-antigen homologs from five different bacterial species: *P. aeruginosa*, *Y. pestis*, *Photorhabdus luminescens*, *Aeromonas salmonicida* and *Vibrio parahaemolyticus* [76]. Abs targeting the T3SS needle tip will likely adopt the secretion blockade mechanism of pathogenesis prevention [34]. This has two variations: translocation blockade or a true secretion blockade (Figure 4). Translocation is defined as secretion directly into a host cell while secretion is an expulsion of protein through the T3SS needle. When an anti-needle tip Ab binds to the needle tip it can create a physical barrier between the tip and the translocon. This barrier prevents secreted effectors from directly entering the host cell and instead are secreted into the extracellular matrix. True secretion blockades occur when the Ab binding prevents the effector proteins from exiting the needle. When designing anti-needle tip Abs, a true blockade style inhibition is desirable because the effector proteins are never released.

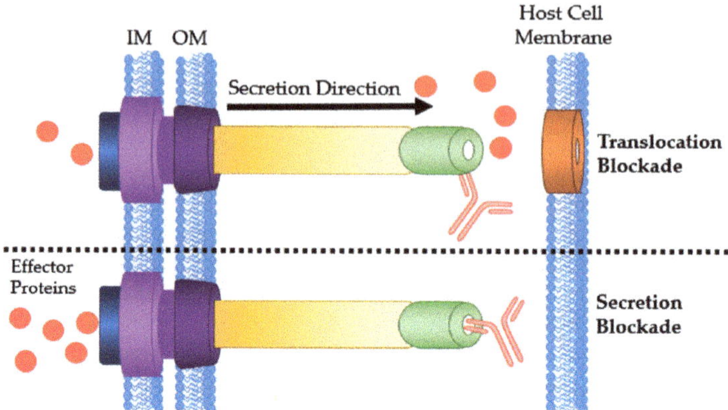

Figure 4. Antibody binding sites on the T3SS needle tip. **Top**: Ab bound to the needle tip protein resulting in a translocation blockade. **Bottom**: Ab bound to the needle tip protein physically blocking any effector secretion. IM: bacterial inner membrane; OM: bacterial outer membrane.

Vaccines for bubonic plague have existed since the late 19th century [77,78]. In 1958, researchers noticed V antigen was present in pathogenic strains of *Yersinia*, but not in non-pathogenic strains [79]. Passive transfer of anti-V antigen antisera limits infection by *Yersinia pestis*, the bacterial agent of the plague [80]. Motin et al. confirmed this immunogenicity by testing mAbs and recombinant Abs against V antigen in 1994 [81]. Once the T3SS was discovered, the V antigen was determined to be LcrV, the needle tip protein of the *Y. pestis* T3SS [82,83]. Abs targeting LcrV were sufficient to prevent translocation of effector proteins by the *Y. pestis* T3SS [84]. Since this time, there has been an explosion of research regarding anti-LcrV antibodies [72,85–90].

Miller et al. demonstrated the importance of cross-strain and cross-species compatibility when designing therapeutic Abs. PAbs and mAbs against one strain of *Y. pestis* LcrV were able to bind the LcrV of two other strains of *Y. enterocolitica*, but no other species or strains tested [85]. Ivanov et al. confirmed that anti-LcrV mAbs were sufficient to directly prevent the secretion of Yop effector proteins. Comparison of IgG mAbs to deglycosylated $F(ab')_2$ and unmodified Fab revealed that the mAb did not require opsonophagocytosis to neutralize Yop translocation [86]. Their work provided a foundation to show that anti-needle tip Abs were acting as a secretion blockade and not a translocation blockade [34].

P. aeruginosa has a needle tip protein that is a homolog to LcrV called PcrV [2]. There is a wide breadth of knowledge on anti-PcrV Abs [61,66,91–101]. The use of anti-PcrV Abs as a vaccine or therapeutic is a topic of interest for many researchers. Taking inspiration from the successes of anti-LcrV therapeutics, Shime et al. investigated anti-PcrV polyclonal IgG and $F(ab')_2$ [61]. Anti-PcrV whole IgG significantly improved the survival rate of mice infected with otherwise lethal doses of *P. aeruginosa* and protected against septic shock in an airspace-infected rabbit model. $F(ab')_2$ derived from the IgG was also tested and the results were comparable.

Polyclonal anti-PcrV IgG as a passive immunization has been evaluated in other models. Burned mouse model results showed that anti-PcrV IgG was significantly better than control IgG at increasing survival rates of mice challenged with a lethal dose of *P. aeruginosa* [91]. Anti-PcrV Abs showed increased effectivity in combination therapy with three separate antibiotics against acute *P. aeruginosa* infection. The combination showed better effectivity than any of the antibiotics or Ab when administered alone [92]. IgY is chicken egg yolk immunoglobulin that is functionally homologous to mammalian IgG. IgY does not react with the mammalian complement system. This reduces the inflammatory response during administration, which makes it an attractive alternative to mammalian IgG [102]. Ranjbar et al. have recently shown that anti-PcrV IgY was more protective against *P. aeruginosa* acute pneumonia and burn-associated infections than control IgY. IgY was comparable to IgG and could serve to be a more affordable alternative in the future [93].

In 2002, Frank et al. tested anti-PcrV mAbs from 80 strains of *P. aeruginosa* to determine which would confer the highest immunoprotection. T3SS secretion assays were performed to determine which mAbs could prevent the translocation of ExoU, a T3SS effector of *P. aeruginosa*. The mAbs showing T3SS inhibition were evaluated for their ability to protect against an otherwise lethal challenge of *P. aeruginosa* in a mouse survival assay. MAb166 was the only antibody tested that showed dose-dependent T3SS inhibition and immunoprotection [94]. De Tavernier et al. has turned to computational methods to find more effective anti-PcrV nanobodies. Three hundred and sixty-one bivalent and biparatopic nanobodies were screened computationally for their ability to cause secretion blockades. T3SS secretion inhibition assays using these nanobodies was performed, followed by mouse survival assays. The most potent nanobody, 13F07-5H01, was effective as a prophylactic up to 24 h after administration, the longest time point tested [95].

Some anti-PcrV mAb therapies have gone into human trials. One of these, termed KB001-A, is a human PEGylated IgG monoclonal anti-PcrV Fab [96–98]. In France, KB001-A underwent phase I and II clinical trials for ventilator-associated *P. aeruginosa* and was considered to be safe and well-tolerated. It did not advance to phase III trials due to a lack of evidence that it reduced pulmonary disease exacerbation in mechanically ventilated patients [103,104]. KB001-A also underwent phase II clinical trials in the US for treatment of chronic pneumonia in cystic fibrosis patients and performed well in safety-based phase I trials but did not continue to phase III [105].

More recently, an alternative anti-PcrV mAb, MEDI3902, has entered human clinical trials. This mAb is bispecific, targeting both PcrV and Psl exopolysaccharide, an anti-biofilm formation target. MEDI3902 showed a dose-dependent survival increase and a decrease in bacterial load in both rabbit and mouse *P. aeruginosa* challenge models. MEDI3902 also reduced lung inflammation caused by bacterial colonization [99]. Le et al. showed MEDI3902 was effective as a treatment and a prophylactic for acute blood and acute lung *P. aeruginosa* infections. Combination therapy with a subtherapeutic dose of the antibiotic meropenem enhanced effectivity [66]. MEDI3902 performed well in phase I clinical trials and is currently undergoing phase IIb trials in the US [106,107].

EspA is the needle tip in the T3SS of *E. coli* [2]. In 2006, recombinant anti-EspA pAbs were shown to reduce actin cytoskeleton rearrangement of the host cell but did not show any reduction of bacterial adhesion [108]. This was the first report of anti-EspA Abs showing inhibitory effects upon the T3SS. Girard et al. investigated the effectivity of bacterial adherence inhibition with IgY to multiple *E. coli* T3SS-related colonization factors, one of which was EspA. Unfortunately, the anti-EspA polyclonal IgY did not significantly reduce bacterial adhesions in multiple strains of pathogenic *E. coli* [109]. The Girard results were disputed when Cook et al. published that anti-EspA IgY and rat IgG reduced adherence of *E. coli* to HeLa cells and prevented T3SS secretion [110].

Yu et al. discovered a novel anti-EspA mAb, 1H10 that provided protection for mice in a survival assay and blocked actin polymerization within host cells [111]. In 2014 Praekelt et al. researched the five major variants of EspA to create over 200 mAbs. Three separate mAbs reacted with multiple EspA variants [112]. While this research was intended to create a better *E. coli* diagnostic test, the results could be applied to treat or prevent *E. coli* infections.

Salmonella enterica serovar Enteritidis (*S.* Enteritidis) causes a large portion of food-related illnesses around the world. *Salmonella* has two T3SSs. The first, called T3SS1, is encoded by the SPI-1 pathogenicity island and is used for host cell entry. The second, called T3SS2, is encoded by the SPI-2 pathogenicity island and is used for further pathogenesis once inside the host cell. SipD is the T3SS1 needle tip protein [2]. Desin et al. have shown that anti-SipD pAbs in sera protected human Caco-2 cells from the entry of *S.* Enteritidis [113].

The needle tip protein in *Shigella* spp. is IpaD [2]. Barta et al. showed that small molecule binding to IpaD induced conformation changes. These changes are accompanied by a significant reduction in the invasive potential of *Shigella* [114]. A panel of anti-SipD nanobodies was tested for their binding affinity and their epitopes were determined. Nanobodies targeting the same section of the needle tip protein resulted in the same conformational change [115]. This research supports the theory that anti-needle tip antibodies may be able to create secretion blockades without physically blocking the effectors from being secreted.

4.2. Translocon

The translocon is made of two proteins that are secreted and subsequently enter the host cell membrane and form a pore. They then link to the needle tip to complete the channel between the pathogen and host cells. Humans naturally produce antibodies against translocon proteins [22]. Abs that target these proteins adopt similar mechanisms to anti-needle tip Abs. They can bind at three time points: before integration into the host membrane, preventing pore formation; after pore formation but before the needle tip has attached, therefore blocking it from attaching and creating a translocation blockade; or once the T3SS is active, mark the cell for ADCC or opsonization (Figure 5).

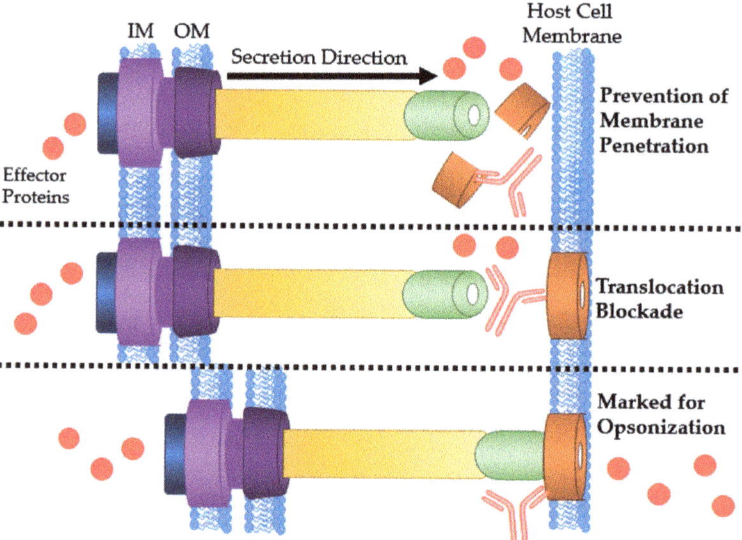

Figure 5. Mechanisms of Ab interaction with T3SS translocon proteins. **Top**: Ab binding to translocon proteins after secretion preventing host membrane penetration and pore formation. **Middle**: Physical blockage of translocation by Ab binding. **Bottom**: Ab bound to the translocon marking it for opsonization or ADCC. IM: bacterial inner membrane; OM: bacterial outer membrane.

The *Y. pestis* T3SS translocon is composed of the proteins YopB & YopD [2]. Ivanov et al. investigated these proteins and their potential for therapeutic use in the treatment and prevention of *Y. pestis* infection. Active and passive immunization protected against otherwise lethal injections of *Y. pestis* in mice [116]. They posited YopD was the dominant immunogen due to the higher titers of Abs and that passive immunization with anti-YopD Abs alone would be enough to confer protection.

The translocon of *E. coli*'s T3SS is formed with two proteins, EspB and EspD [2]. A study of Brazilian children showed that those with naturally occurring anti-EspB Abs were less likely to have a severe EHEC infection [117]. Maternal vaccination with EspB affords passive immunization of offspring with anti-EspB Abs. These Abs are effective at reducing risk for *E. coli* infection and increasing infection survival rates [24]. Similar experiments using mice have shown comparable survival results. Placebo-vaccinated mothers had offspring with significantly higher plasma urea concentrations, a marker of renal failure [118].

Salmonella Enteritidis' translocon is comprised of proteins SipB and SipC [2]. Although there is limited information about antibodies targeting these proteins there is evidence of a mAb indirectly preventing the formation of the T3SS-1 translocon in *Salmonella*. The mAb Sal4 targets a surface polysaccharide of *Salmonella* named O antigen. Forbes et al. observed that Sal4 appeared to interfere with flagellum-based motility and T3SS-mediated entry of the host intestinal epithelium [119]. Interference with host cell entry by Sal4 was due to inhibition of the T3SS. Direct interaction of SipB or SipC with Sal4 was not tested, but other T3SS components and effectors were eliminated as antigens for Sal4 [120].

4.3. Basal Body

The basal body of the T3SS contains four major components. These include an ATPase that powers secretion, the lower ring within the inner bacterial membrane, an export apparatus that is visible between the bacterial membranes, and the upper ring located in the outer membrane of the bacterial cell. Of these components, only the upper ring of the basal body is exposed to the extracellular matrix. Research regarding therapeutic anti-upper ring Abs has focused on Abs that will mark the T3SS, and therefore the pathogen, for opsonization [121]. There is a possibility that the needle formation could be inhibited if Abs bind to the correct area of the upper ring (Figure 6). YscC makes up the upper ring of the *Y. pestis* T3SS basal body [2]. Goodin et al. have shown that passive immunization with anti-YscC pAbs induced the mouse immune system to produce more anti-YscC Abs but did not provide sufficient protection against a lethal *Y. pestis* challenge in comparison to an F1 & LcrV protein-based vaccine [121]. The outer membrane ring of the T3SS basal body shares high similarity to other outer membrane proteins in secretion systems unrelated to pathogenesis [122]. The similarity could, in theory, allow for Ab binding to secretion systems on commensal bacteria. This potentially reduced specificity along with the lack of protection in the Goodin study highlights the challenge of using the basal body as a target for therapeutic Abs.

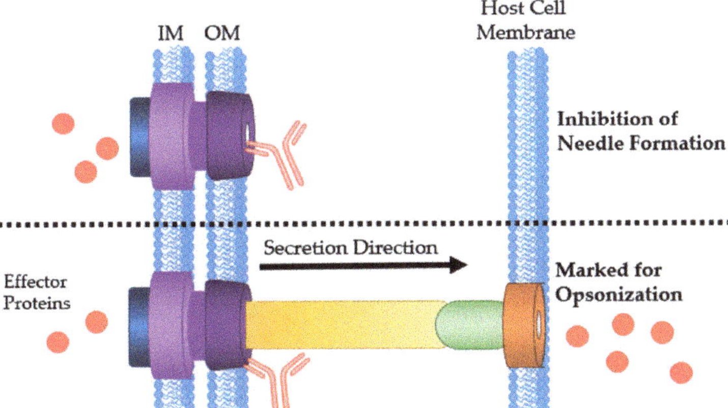

Figure 6. Mechanisms of Ab interaction with the T3SS basal body. **Top**: Ab bound to the outer membrane ring of the T3SS basal body preventing needle formation. **Bottom**: Ab bound to the basal body marking it for opsonization or ADCC. IM: bacterial inner membrane; OM: bacterial outer membrane.

4.4. Effector Proteins

The critical function of the T3SS is to secrete proteins directly into host cells. These effector proteins have a wide variety of mechanisms and purposes [6–9]. Knockout or mutations of some effectors can result in attenuation of virulence and reduced pathogenesis, making them attractive targets [123].

The majority of effector proteins are translocated into the host cell and are unable to leave. This creates a challenge in using antibodies against these particular effectors [56]. Some effectors, however, can transverse and exit the host cell after translocation, making them available for neutralization by antibodies. Other effectors may be presented on the surface of host cells. These proteins are more accessible for targeting by antibodies.

4.4.1. Antibodies Targeting Extracellularly Available Effectors

Yersinia outer proteins (Yops) are a class of T3SS effectors in *Y. pestis*, *Y. enterocolitica* and *Y. pseudotuberculosis*. Some Yops are present both internally and in the extracellular space (Figure 7, Left). Akopyan et al. observed the presence of Yops outside host cells and found that YopE is localized to the surface of host cells, but not necessarily where the T3SS is attached. They hypothesized that some Yops (e.g., YopE) can enter host cells in a T3SS-independent manner by hijacking host transporters while others must utilize the T3SS (e.g., YopH) [124].

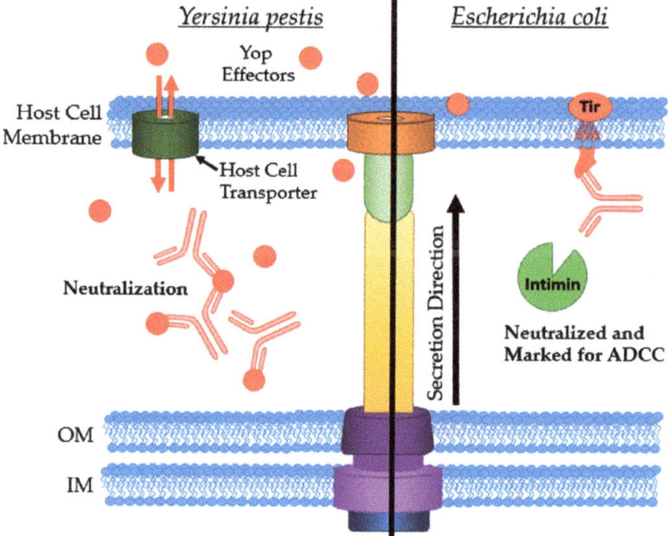

Figure 7. Antibodies targeting extracellularly available effector proteins. **Left**: Yop effectors being neutralized when in the extracellular matrix. **Right**: Ab bound to Tir marking the host cell for ADCC and neutralizing Tir by preventing intimin binding. IM: bacterial inner membrane; OM: bacterial outer membrane.

Ivanov et al. discovered that anti-YopE Abs inhibited bacterial infection but to a lesser extent than Abs against the translocon proteins, YopB, and YopD [116]. Later, Singh et al. vaccinated mice against rVE, a YopE-LcrV fusion protein, using both active and passive approaches. Active immunization with the protein conferred high titers of Abs against both antigens. The serum from vaccinated mice was used to vaccinate another batch of mice. These passively immunized mice showed nearly 90% survival against a lethal challenge of *Y. enterocolitica* with the majority showing no signs or symptoms of infection even after necropsy [125]. YopE specific $CD8^+$ T cells are naturally occurring immune cells coated in Abs that recognize YopE. Immunization of mice with these cells was 60% protective against mucosal and systemic *Y. pseudotuberculosis* infection in a survival assay. YopE specific $CD8^+$ T cells, not just anti-YopE Abs, are required for protection against infection [126].

The essential effector YopM is a modulator of kinases PRK1 and PRK2. This modulation eventually leads to a reduction in pro-inflammatory cytokines lessening the effectivity of the host immune response [8].

YopM is the first effector to be recognized as a bacterial cell-penetrating protein as it can leave and re-enter host cells after translocation [127]. Neutralization of YopM would in theory restore the host cytokine response as well as precipitation of YopM, resulting in the recruitment of phagocytes to the infection site (Figure 7, Left). Rüter et al. have isolated an anti-YopM mAb that binds to multiple strains of pathogenic *Yersinia*'s YopM but not to YopM from non-pathogenic strains [128]. This specificity may be beneficial when designing therapeutics and diagnostics.

The translocated intimin receptor (Tir) is one of the first effector proteins translocated by *E. coli*'s T3SS into host cells. After folding, Tir integrates into the host membrane to provide a pedestal for adhesion via intimin binding [129]. This extracellular expression allows for Abs to bind without having to cross the eukaryotic membrane (Figure 7, Right). Girard et al. found that anti-Tir IgY were effective at preventing bacterial adhesion to host cells in a porcine model against both porcine and human strains of enteropathogenic *E. coli* (EPEC) [109]. Ruano-Gallego et al. assessed the potential of an anti-Tir nanobody, TD4, as a potential treatment or prophylactic for EHEC infections. TD4 inhibits the attachment of EHEC to HeLa cells and reduces adherence to human colonic mucosa [130]. Anti-intimin IgY was effective at reducing adherence of both EPEC strains to host gut epithelial tissue in an ileal loop assay as well as in oral administration of the IgY [109]. Kühne et al. purified nanobodies binding to the Tir-binding domain of intimin [131]. Saberianfar et al. investigated the Tir-binding domain of intimin as a target to isolate sdAbs from tobacco leaves. These sdAbs were used to design a chimeric Ab, V_HH10-IgA. This Ab inhibited four strains of EHEC from adhering to host cells, with three of the four completely inhibited [74]. V_HH10-IgA's cross-serotype inhibition of bacterial adhesion is highly promising for future studies.

4.4.2. Adjuvating Antibodies Targeting Effectors

Effector proteins that are only present within the host cell become available upon cell lysis [25]. Abs targeting these effectors will not be able to prevent T3SS formation or secretion but will add to the host immune response [132,133]. Some infected host cells will also participate in antigen presentation. This is the process of breaking up non-native proteins, such as T3SS effectors, and displaying the fragments on the host membrane surface so that Abs can access them [25]. Although not as immunoprotective as T3SS inhibitory Abs, adjuvating Abs can be important to increase natural host immune response. Their use is common in combination with inhibitory Abs or as diagnostics.

Desin et al. were inspired by traditional research on T3SS protein-based vaccines in cattle and other ruminants, the main animal reservoir for Shiga toxin-producing *E. coli* (STEC) [18]. Passive immunization using uncharacterized rabbit-produced sera was sufficient to block adherence of *E. coli* to host cells [134]. Desin et al. also tested antisera containing pAbs against three effectors: Tir, EspF, and NleA (EspI), along with the needle tip protein, EspA, and a translocon component, EpB. EspF assists in host cytoskeleton rearrangement and inhibits host cell apoptosis. NleA localizes to the Golgi apparatus in the host cell and disturbs ER to Golgi transport. Desin et al. showed Abs against either EspF or NleA inhibited bacterial adherence of two STEC strains ($STEC_{O103}$ & $STEC_{O157}$).

ExoS, exoenzyme S, is an effector protein secreted by the *P. aeruginosa* T3SS. ExoS, along with three other effectors: ExoU, ExoY, and ExoT, assist in the prevention of wound repair in the host by reducing the immune response and causing damage to host mucosal membranes [135]. Knockout and mutations of ExoS result in reduced pathogenesis of *P. aeruginosa* [123]. Corech et al. examined the serum of patients with *P. aeruginosa* infections and found they universally had significant titers of IgG against PopB, PcrV, and ExoS [136]. These antibodies may be viable candidates for pharmaceutical development.

IncA is an effector secreted by the *C. trachomatis* T3SS and generates robust IgG responses in humans. *C. trachomatis* enters host cells where it replicates within vacuoles called inclusions. IncA has a role in the homotypic fusion of these inclusions. Pathogenic *C. trachomatis* has attenuated host cell invasion in the presence of anti-IncA Abs [137]. Tsai et al. sequenced IncA from multiple *C. trachomatis* isolates and found that they were nearly identical to all human serotypes sequenced thus far, suggesting that anti-IncA Abs should react with IncA from multiple serotypes. Anti-IncA Abs were found in 52% of urine samples and 71% of genital samples from *C. trachomatis*-infected patients [138]. Although further research is needed to confirm the immunogenicity of IncA, these results support the use of anti-IncA Abs as diagnostic or therapeutic antibodies.

An effector secreted by the *Salmonella* T3SS2 is SpiC, also called SsaB. SpiC interferes with host cell trafficking and knockouts show attenuated virulence and decreased T3SS2 activity [139]. Geng et al. developed seven anti-SpiC mAbs. These mAbs bound specifically to SpiC and not to the His or GST, both of which were used in the mAb isolation process [140]. Immunogenicity data for these mAbs was not presented, but there is potential for their use as therapeutics or diagnostic tools.

4.4.3. Antibodies Targeting Intracellular Effectors and Transcription Factors

Delivery of antibodies into cells is required to access intracellular targets. One method to overcome issues of cell penetration is to express the antibody within the cell. Intrabodies are internally expressed antibodies. Gene transport mechanisms are used to deliver the DNA encoding the therapeutic antibody or antibody fragment inside the cell where it can be transcribed (Figure 8A) [31,32,141]. Another method of internalizing Abs is to pair them with a membrane-penetrating peptide (MPP). Attaching an MPP to an antibody or antibody fragment allows for the therapeutic antibody or fragment to physically transverse the membrane. The MPP destabilizes bacterial cell membranes and enables the fused protein to traverse the outer membrane (Figure 8B) [142].

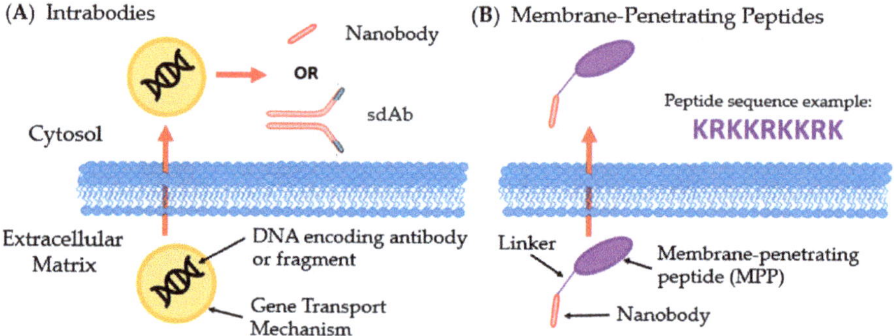

Figure 8. Innovative methods for intracellular delivery of Abs. (**A**) Intrabodies enter the host cell containing DNA encoding a therapeutic Ab or antibody fragment, most often a sdAb or nanobody; (**B**) Membrane-penetrating peptides are linked to Abs or fragments, commonly nanobodies, and allows for transversing membranes.

SpvB is a cytotoxic effector secreted through the *S.* Typhimurium T3SS2 into the host cell from the vacuole containing the pathogen. Once in the host, it catalyzes ADP-ribosylation of actin and eventually causes host cell apoptosis. Alzogaray et al. have developed a nanobody that binds with a high affinity to SpvB. The nanobodies were expressed as intrabodies to neutralize the effector within the host cell [143]. The anti-SpvB antibodies stopped the action of SpvB in an ATP-induced actin polymerization fluorescence-based assay in vitro and in RAW macrophages [143].

Targeting a transcriptional regulator of the T3SS could prevent the proteins that make up the T3SS from being produced in the first place [144]. This method of pathogenesis prevention is not common, as the antibody in question would have to enter the pathogen rather than a eukaryotic host cell. SpuE is a transcriptional regulator of the T3SS in *P. aeruginosa* and regulates the expression of ExsA, a master regulator of the T3SS [2], via inhibition of the exsCEBA promoter. Zhang et al. derived an anti-SpuE nanobody fused with a membrane-penetrating peptide (scFv5-MPP) to assist in delivering the nanobody into the bacterial cell [145]. scFv5-MPP allosterically inhibits the expression of the T3SS and attenuates virulence of *P. aeruginosa* in a *C. elegans* animal gut infection model. X-ray crystallography and molecular dynamics simulations were used to observe conformational changes upon antibody binding to SpuE. This conformational change may cause a reduction in spermidine uptake by *P. aeruginosa* leading to attenuation of virulence similar to physical neutralization [145]. A similar antibody, Mab 4E4, protects A549 cells against *P. aeruginosa* infection by reducing T3SS expression and polyamine uptake. Wang et al. have recently tested Mab 4E4 in vivo and found that single injection vaccination with the antibody significantly increase survival rates of mice given a four-fold lethal dose of *P. aeruginosa* infection and protected them from severe alveolar destruction [146]. These examples validate the strategy of anti-transcriptional regulator antibodies as therapeutics or prophylactics against T3SS-utilizing bacteria.

5. Conclusions

A diverse array of antibodies has been used to inhibit the T3SS. These Abs bind to proteins of the injectisome including the needle tip, translocon, basal body, and effectors. Transcriptional regulators of the T3SS have also been targeted to prevent the formation of the T3SS, but as they are intracellular targets and require innovative cell-penetrating Abs. These can include entrapping DNA encoding the Ab in intrabodies or attachment of membrane-penetrating proteins to the Ab. Once the Ab has reached its target there are multiple mechanisms it can employ to attenuate virulence or increase the host immune response. In general, these Abs neutralize the effectors, mark the bacterial cell for phagocytes to attack, or mark the infected host cell for ADCC by NK cells. Sometimes Abs adopt more specific mechanisms. For example, when targeting the needle tip or translocon the Ab can physically block the secretion of effectors in a secretion blockade. Several anti-T3SS mAbs have advanced to clinical trials, but none have yet made it to market. As we learn more about how these antibodies function there will undoubtedly be potential for improvement of their therapeutic effects, cost of production, and the ease of their delivery.

Author Contributions: Writing—original draft preparation, J.A.H. and A.E.M.; writing—review and editing, J.A.H. and A.E.M. All authors have read and agreed to the published version of the manuscript

Funding: This research was funded by startup funds from the School of Pharmacy at Virginia Commonwealth University (VCU). This work was also supported by VCU's CTSA (UL1TR002649 from the National Center for Advancing Translational Sciences) and the CCTR Endowment Fund of Virginia Commonwealth University.

Acknowledgments: We thank Heather A. Pendergrass for helpful discussions.

Conflicts of Interest: The authors declare no conflict of interest.

References

1. Lara-Tejero, M.; Galán, J.E. The injectisome, a complex nanomachine for protein injection into mammalian cells. *EcoSal Plus* **2019**, *8*, 245–259. [CrossRef]
2. Hu, Y.; Huang, H.; Cheng, X.; Shu, X.; White, A.P.; Stavrinides, J.; Köster, W.; Zhu, G.; Zhao, Z.; Wang, Y. A global survey of bacterial type III secretion systems and their effectors. *Environ. Microbiol.* **2017**, *19*, 3879–3895. [CrossRef] [PubMed]
3. Pendergrass, H.A.; May, A.E. Natural product type III secretion system inhibitors. *Antibiotics* **2019**, *8*, 162. [CrossRef] [PubMed]
4. Cornelis, G.R.; Van Gijsegem, F. Assembly and function of type III secretory systems. *Annu. Rev. Microbiol.* **2000**, *54*, 735–774. [CrossRef] [PubMed]
5. Cheung, M.; Shen, D.K.; Makino, F.; Kato, T.; Roehrich, A.D.; Martinez-Argudo, I.; Walker, M.L.; Murillo, I.; Liu, X.; Pain, M.; et al. Three-dimensional electron microscopy reconstruction and cysteine-mediated crosslinking provide a model of the type III secretion system needle tip complex. *Mol. Microbiol.* **2015**, *95*, 31–50. [CrossRef] [PubMed]
6. Hume, P.J.; Singh, V.; Davidson, A.C.; Koronakis, V. Swiss army pathogen: The *Salmonella* entry toolkit. *Front. Cell Infect. Microbiol.* **2017**, *7*, 348. [CrossRef] [PubMed]
7. Mattock, E.; Blocker, A.J. How do the virulence factors of *Shigella* work together to cause disease? *Front. Cell Infect. Microbiol.* **2017**, *7*, 64. [CrossRef]
8. Zhang, L.; Mei, M.; Yu, C.; Shen, W.; Ma, L.; He, J.; Yi, L. The functions of effector proteins in *Yersinia* virulence. *Pol. J. Microbiol.* **2016**, *65*, 5–12. [CrossRef]

9. Morrow, K.A.; Ochoa, C.D.; Balczon, R.; Zhou, C.; Cauthen, L.; Alexeyev, M.; Schmalzer, K.M.; Frank, D.W.; Stevens, T. *Pseudomonas aeruginosa* exoenzymes U and Y induce a transmissible endothelial proteinopathy. *Am. J. Physiol. Lung Cell Mol. Physiol.* **2016**, *310*, L337–L353. [CrossRef]
10. Marshall, N.C.; Brett Finlay, B. Targeting the type III secretion system to treat bacterial infections. *Expert Opin. Ther. Targets* **2014**, *18*, 137–152. [CrossRef]
11. Matsuda, S.; Okada, R.; Tandhavanant, S.; Hiyoshi, H.; Gotoh, K.; Iida, T.; Kodama, T. Export of a *Vibrio parahaemolyticus* toxin by the Sec and type III secretion machineries in tandem. *Nat. Microbiol.* **2019**, *4*, 781–788. [CrossRef] [PubMed]
12. Stone, C.B.; Bulir, D.C.; Emdin, C.A.; Pirie, R.M.; Porfilio, E.A.; Slootstra, J.W.; Mahony, J.B. *Chlamydia pneumoniae* CdsL regulates CdsN ATPase activity, and disruption with a peptide mimetic prevents bacterial invasion. *Front. Microbiol.* **2011**, *2*, 21. [CrossRef] [PubMed]
13. Nakamura, K.; Shinoda, N.; Hiramatsu, Y.; Ohnishi, S.; Kamitani, S.; Ogura, Y.; Hayashi, T. Horiguchi YBspR/BtrA, an anti-σ, factor. Regulates the ability of *Bordetella bronchiseptica* to cause cough in rats. *Msphere* **2019**, *4*, e00093. [CrossRef] [PubMed]
14. Berube, B.J.; Murphy, K.R.; Torhan, M.C.; Bowlin, N.O.; Williams, J.D.; Bowlin, T.L.; Moir, D.T.; Hauser, A.R. Impact of type III secretion effectors and of phenoxyacetamide inhibitors of type III Secretion on abscess formation in a mouse model of *Pseudomonas aeruginosa* infection. *Antimicrob. Agents Chemother.* **2017**, *61*, e01202. [CrossRef] [PubMed]
15. Duncan, M.C.; Linington, R.G.; Auerbuch, V. Chemical inhibitors of the type three secretion system: Disarming bacterial pathogens. *Antimicrob. Agents Chemother.* **2012**, *56*, 5433–5441. [CrossRef]
16. Kolár, M.; Urbánek, K.; Látal, T. Antibiotic selective pressure and development of bacterial resistance. *Int. J. Antimicrob. Agents* **2001**, *17*, 357–363. [CrossRef]
17. Kimura, K.; Iwatsuki, M.; Nagai, T.; Matsumoto, A.; Takahashi, Y.; Shiomi, K.; Omura, S.; Abe, A. A small-molecule inhibitor of the bacterial type III secretion system protects against in vivo infection with *Citrobacter rodentium*. *J. Antibiot.* **2011**, *64*, 197–203. [CrossRef]
18. Loureiro, I.; Frankel, G.; Adu-Bobie, J.; Dougan, G.; Trabulsi, L.R.; Carneiro-Sampaio, M.M.S. Human colostrum contains IgA antibodies reactive to enteropathogenic *Escherichia coli* virulence-associated proteins: Intimin, BfpA, EspA, and EspB. *J. Pediatr. Gastroenterol. Nutr.* **1998**, *27*, 166–171. [CrossRef]
19. Shimanovich, A.A.; Buskirk, A.D.; Heine, S.J.; Blackwelder, W.C.; Wahid, R.; Kotloff, K.L.; Pasetti, M.F. Functional and antigen-specific serum antibody levels as correlates of protection against shigellosis in a controlled human challenge study. *Clin. Vaccine Immunol.* **2017**, *24*, e00412. [CrossRef]
20. Gavilanes-Parra, S.; Mendoza-Hernández, G.; Chávez-Berrocal, M.E.; Girón, J.A.; Orozco-Hoyuela, G.; Manjarrez-Hernández, A. Identification of secretory immunoglobulin A antibody targets from human milk in cultured cells infected with enteropathogenic *Escherichia coli* (EPEC). *Microb. Pathog.* **2013**, *64*, 48–56. [CrossRef]
21. Li, Y.; Frey, E.; Mackenzie, A.M.R.; Finlay, B.B. Human response to *Escherichia coli* O157:H7 infection: Antibodies to secreted virulence factors. *Infect. Immun.* **2000**, *68*, 5090–5095. [CrossRef] [PubMed]
22. Durand, D.; Ochoa, T.J.; Bellomo, S.M.E.; Contreras, C.A.; Bustamante, V.H.; Ruiz, J.; Cleary, T.G. Detection of secretory immunoglobulin a in human colostrum as mucosal immune response against proteins of the type III secretion system of *Salmonella*, *Shigella* and enteropathogenic *Escherichia coli*. *Pediatr. Infect. Dis. J.* **2013**, *32*, 1122–1126. [CrossRef]
23. Rabinovitz, B.C.; Gerhardt, E.; Tironi Farinati, C.; Abdala, A.; Galarza, R.; Vilte, D.A.; Ibarra, C.; Cataldi, A.; Mercado, E.C. Vaccination of pregnant cows with EspA, EspB, γ-intimin, and Shiga toxin 2 proteins from *Escherichia coli* O157:H7 induces high levels of specific colostral antibodies that are transferred to newborn calves. *J. Dairy Sci.* **2012**, *95*, 3318–3326. [CrossRef] [PubMed]
24. Rabinovitz, B.C.; Vilte, D.A.; Larzábal, M.; Abdala, A.; Galarza, R.; Zotta, E.; Ibarra, C.; Mercado, E.C.; Cataldi, A. Physiopathological effects of *Escherichia coli* O157: H7 inoculation in weaned calves fed with colostrum containing antibodies to EspB and Intimin. *Vaccine* **2014**, *32*, 3823–3829. [CrossRef] [PubMed]
25. Mayne, E.; Prinz, W.; Van Dixhoorn, M.S.; Mayne, E.; Wadee, A.A. Immunology. In *Molecular Medicine for Clinicians*; Mendelow, B., Ramsay, M.N., Chetty, W.S., Eds.; Wits University Press: Johannesburg, South Africa, 2009; pp. 289–310.

26. Davies, D.H. Antigen Discovery for Vaccines Using High-throughput Proteomic Screening Techniques. In *Vaccinology: Principles and Practice*; Morrow, W.J.W., Sheikh, N.A., Schmidt, C.S., Davies, H.D., Eds.; Blackwell Publishing Ltd.: Oxford, UK, 2012; pp. 150–167.
27. Li, Y.; Jin, L.; Chen, T. The effects of secretory IgA in the mucosal immune system. *BioMed Res. Int.* **2020**, *2020*, 2032057. [CrossRef] [PubMed]
28. Yanaka, S.; Yogo, R.; Kato, K. Biophysical characterization of dynamic structures of immunoglobulin G. *Biophys. Rev.* **2020**, *12*, 637–645. [CrossRef] [PubMed]
29. Zurawski, D.V.; Mclendon, M.K. Monoclonal antibodies as an antibacterial approach against bacterial pathogens. *Antibiotics* **2020**, *9*, 155. [CrossRef]
30. Nagy, E.; Nagy, G.; Power, C.A.; Badarau, A.; Szijártó, V. Anti-bacterial monoclonal antibodies. *Adv. Exp. Med. Biol.* **2017**, *1053*, 119–153.
31. Hu, Y.; Liu, C.; Muyldermans, S. Nanobody-based delivery systems for diagnosis and targeted tumor therapy. *Front. Immunol.* **2017**, *8*, 1442. [CrossRef]
32. Kolkman, J.A.; Law, D.A. Nanobodies-From llamas to therapeutic proteins. *Drug Discov. Today Technol.* **2010**, *7*, e139–e146. [CrossRef]
33. Forthal, D.N. Functions of antibodies. *Microbiol. Spectr.* **2014**, *2*, 1–17. [PubMed]
34. Sawa, T.; Kinoshita, M.; Inoue, K.; Ohara, J.; Moriyama, K. Immunoglobulin for treating bacterial infections: One more mechanism of action. *Antibodies* **2019**, *8*, 52. [CrossRef]
35. Goldberg, B.S.; Ackerman, M.E. Antibody-mediated complement activation in pathology and protection. *Immunol. Cell Biol.* **2020**, *98*, 305–317. [CrossRef] [PubMed]
36. Abramov, V.M.; Kosarev, I.V.; Motin, V.L.; Khlebnikov, V.S.; Vasilenko, R.N.; Sakulin, V.K.; Machulin, A.V.; Uversky, V.N.; Karlyshev, A.V. Binding of LcrV protein from *Yersinia pestis* to human T-cells induces apoptosis, which is completely blocked by specific antibodies. *Int. J. Biol. Macromol.* **2019**, *122*, 1062–1070. [CrossRef] [PubMed]
37. Jenner, E. On the origin of the vaccine inoculation. *Med. Phys. J.* **1801**, *5*, 505–508. [PubMed]
38. Davies, D.H.; Schmidt, C.S.; Sheikh, N.A. Concept and Scope of Modern Vaccines. In *Vaccinology: Principles and Practice*; Morrow, W.J.W., Sheikh, N.A., Schmidt, C.S., Davies, H.D., Eds.; Blackwell Publishing Ltd.: Oxford, UK, 2012; pp. 3–11.
39. Frey, S.E.; Lottenbach, K.; Graham, I.; Anderson, E.; Bajwa, K.; May, R.C.; Mizel, S.B.; Graff, A.; Belshe, R.B. A phase I safety and immunogenicity dose escalation trial of plague vaccine, Flagellin/F1/V, in healthy adult volunteers (DMID 08-0066). *Vaccine* **2017**, *35*, 6759–6765. [CrossRef]
40. Boros, P.; Gondolesi, G.; Bromberg, J.S. High dose intravenous immunoglobulin treatment: Mechanisms of action. *Liver Transpl.* **2005**, *11*, 1469–1480. [CrossRef]
41. Afonso, A.F.B.; João, C.M.P. The production processes and biological effects of intravenous immunoglobulin. *Biomolecules* **2016**, *6*, 15. [CrossRef]
42. Lee, J.L.; Mohamed Shah, N.; Makmor-Bakry, M.; Islahudin, F.H.; Alias, H.; Noh, L.M.; Mohd Saffian, S. A systematic review and meta-regression analysis on the impact of increasing IgG trough level on infection rates in primary immunodeficiency patients on intravenous IgG therapy. *J. Clin. Immunol.* **2020**, *40*, 682–698. [CrossRef]
43. McCusker, C.; Warrington, R. Primary immunodeficiency. *Allergy Asthma Clin. Immunol.* **2011**, *7* (Suppl. 1), S11. [CrossRef]
44. Chaigne, B.; Mouthon, L. Mechanisms of action of intravenous immunoglobulin. *Transfus. Apheresis Sci.* **2017**, *56*, 45–49. [CrossRef] [PubMed]
45. Negi, V.S.; Elluru, S.; Sibéril, S.; Graff-Dubois, S.; Mouthon, L.; Kazatchkine, M.D.; Lacroix-Desmazes, S.; Bayry, J.; Kaveri, S.V. Intravenous immunoglobulin: An update on the clinical use and mechanisms of action. *J. Clin. Immunol.* **2007**, *27*, 233–245. [CrossRef] [PubMed]
46. Ramus, B.; Benbrahim, O.; Chérin, P. Use of intravenous and subcutaneous human immunoglobulins. *Soins Rev. Ref. Infirm.* **2019**, *64*, 13–18.
47. Sriaroon, P.; Ballow, M. Immunoglobulin replacement therapy for primary immunodeficiency. *Immunol. Allergy Clin. N. Am.* **2015**, *35*, 713–730. [CrossRef] [PubMed]
48. Aubron, C.; Berteau, F.; Sparrow, R.L. Intravenous immunoglobulin for adjunctive treatment of severe infections in ICUs. *Curr. Opin. Crit. Care* **2019**, *25*, 417–422. [CrossRef]

49. Kaplon, H.; Muralidharan, M.; Schneider, Z.; Reichert, J.M. Antibodies to watch in 2020. *MAbs* **2020**, *12*, 1703531. [CrossRef]
50. Sécher, T.; Dalonneau, E.; Ferreira, M.; Parent, C.; Azzopardi, N.; Paintaud, G.; Si-Tahar, M.; Heuzé-Vourc'h, N. In a murine model of acute lung infection, airway administration of a therapeutic antibody confers greater protection than parenteral administration. *J. Control Release* **2019**, *303*, 24–33. [CrossRef]
51. Ndungo, E.; Randall, A.; Hazen, T.H.; Kania, D.A.; Trappl-Kimmons, K.; Liang, X.; Barry, E.M.; Kotloff, K.L.; Chakraborty, S.; Mani, S.; et al. A novel *Shigella* proteome microarray discriminates targets of human antibody reactivity following oral vaccination and experimental challenge. *Msphere* **2018**, *3*, e00260. [CrossRef]
52. Hill, J.; Eyles, J.E.; Elvin, S.J.; Healey, G.D.; Lukaszewski, R.A.; Titball, R.W. Administration of antibody to the lung protects mice against pneumonic plague. *Infect. Immun.* **2006**, *74*, 3068–3070. [CrossRef]
53. Clemens, J.; Elyazeed, R.A.; Rao, M.; Savarino, S.; Morsy, B.Z.; Kim, Y.; Wierzba, T.; Naficy, A.; Lee, Y.J. Early initiation of breastfeeding and the risk of infant diarrhea in rural Egypt. *Pediatrics* **1999**, *104*, e3. [CrossRef]
54. Kapil, P.; Papin, J.F.; Wolf, R.F.; Zimmerman, L.I.; Wagner, L.D.; Merkel, T.J. Maternal vaccination with a monocomponent pertussis toxoid vaccine is sufficient to protect infants in a baboon model of Whooping cough. *J. Infect. Dis.* **2018**, *217*, 1231–1236. [CrossRef] [PubMed]
55. Kaplon, H.; Reichert, J.M. Antibodies to watch in 2019. *MAbs* **2019**, *11*, 219–238. [CrossRef] [PubMed]
56. Gura, T. Magic bullets hit the target. *Nature* **2002**, *417*, 584–586. [CrossRef] [PubMed]
57. Hollowell, P.; Li, Z.; Hu, X.; Ruane, S.; Kalonia, C.; Van der Walle, C.F.; Lu, J.R. Recent advances in studying interfacial adsorption of bioengineered monoclonal antibodies. *Molecules* **2020**, *25*, 2047. [CrossRef] [PubMed]
58. Buyel, J.F.; Twyman, R.M.; Fischer, R. Very-large-scale production of antibodies in plants: The biologization of manufacturing. *Biotechnol. Adv.* **2017**, *35*, 458–465. [CrossRef]
59. Gómez-Mantilla, J.D.; Trocóniz, I.F.; Parra-Guillén, Z.; Garrido, M.J. Review on modeling anti-antibody responses to monoclonal antibodies. *J. Pharmacokinet. Pharmacodyn.* **2014**, *41*, 523–536. [CrossRef]
60. Glück, D.; Wiedeck, H.; van Wickern, M.; Wölpl, A.; Northoff, H.; Ahnefeld, F.W.; Grünert, A.; Kubanek, B. Anti-lipopolysaccharide-immunoglobulin (IgG-Anti-LPS) therapy in intensive care patients following surgery from infectious disease. *Infusiontherapie.* **1990**, *17*, 220–223. [CrossRef]
61. Shime, N.; Sawa, T.; Fujimoto, J.; Faure, K.; Allmond, L.R.; Karaca, T.; Swanson, B.L.; Spack, E.G.; Wiener-Kronish, J.P. Therapeutic administration of anti-PcrV F(ab′)$_2$ in sepsis associated with *Pseudomonas aeruginosa*. *J. Immunol.* **2001**, *167*, 5880–5886. [CrossRef]
62. Fasciano, A.C.; Shaban, L.; Mecsas, J.; Alyssa, C. Fasciano1, Lamyaa Shaban2, and J.M.; Fasciano, A.C.; Shaban, L.; Mecsas, J. Promises and challenges of the type three secretion system- injectisome as an anti-virulence target. *EcoSal Plus* **2019**, *8*, 261–276. [CrossRef]
63. Baron, C.; Coombes, B. Targeting bacterial secretion systems: Benefits of disarmament in the microcosm. *Infect. Disord. Drug Targets* **2008**, *7*, 19–27. [CrossRef]
64. Hilf, M.; Yu, V.L.; Sharp, J.; Zuravleff, J.J.; Korvick, J.A.; Muder, R.R. Antibiotic therapy for *Pseudomonas aeruginosa* bacteremia: Outcome correlations in a prospective study of 200 patients. *Am. J. Med.* **1989**, *87*, 540–546. [CrossRef]
65. Secher, T.; Fas, S.; Fauconnier, L.; Mathieu, M.; Rutschi, O.; Ryffel, B.; Rudolf, M. The anti-*Pseudomonas aeruginosa* antibody Panobacumab is efficacious on acute pneumonia in neutropenic mice and has additive effects with meropenem. *PLoS ONE* **2013**, *8*, e73396. [CrossRef] [PubMed]
66. Le, H.N.; Tran, V.G.; Vu, T.T.T.; Gras, E.; Le, V.T.M.; Pinheiro, M.G.; Aguiar-Alves, F.; Schneider-Smith, E.; Carter, H.C.; Sellman, B.R.; et al. Treatment efficacy of MEDI3902 in *Pseudomonas aeruginosa* bloodstream infection and acute pneumonia rabbit models. *Antimicrob. Agents Chemother.* **2019**, *63*, e00710. [CrossRef] [PubMed]
67. Kunert, R.; Reinhart, D. Advances in recombinant antibody manufacturing. *Appl. Microbiol. Biotechnol.* **2016**, *100*, 3451–3461. [CrossRef] [PubMed]
68. Kang, S.; Ren, D.; Xiao, G.; Daris, K.; Buck, L.; Enyenihi, A.A.; Zubarev, R.; Bondarenko, P.V.; Deshpande, R. Cell line profiling to improve monoclonal antibody production. *Biotechnol. Bioeng.* **2014**, *111*, 748–760. [CrossRef]
69. Huang, Y.M.; Hu, W.W.; Rustandi, E.; Chang, K.; Yusuf-Makagiansar, H.; Ryll, T. Maximizing productivity of CHO cell-based fed-batch culture using chemically defined media conditions and typical manufacturing equipment. *Biotechnol. Prog.* **2010**, *26*, 1400–1410. [CrossRef] [PubMed]

70. Itoh, Y.; Ueda, H.; Suzuki, E. Overexpression of bcl-2, apoptosis suppressing gene: Prolonged viable culture period of hybridoma and enhanced antibody production. *Biotechnol. Bioeng.* **1995**, *48*, 118–122. [CrossRef]
71. Nishimiya, D.; Mano, T.; Miyadai, K.; Yoshida, H.; Takahashi, T. Overexpression of CHOP alone and in combination with chaperones is effective in improving antibody production in mammalian cells. *Appl. Microbiol. Biotechnol.* **2013**, *97*, 2531–2539. [CrossRef]
72. Sittner, A.; Mechaly, A.; Vitner, E.; Aftalion, M.; Levy, Y.; Levy, H.; Mamroud, E.; Fisher, M. Improved production of monoclonal antibodies against the LcrV antigen of *Yersinia pestis* using FACS-aided hybridoma selection. *J. Biol. Methods* **2018**, *5*, e100. [CrossRef]
73. Zhou, Y.; Liu, P.; Gan, Y.; Sandoval, W.; Katakam, A.K.; Reichelt, M.; Rangell, L.; Reilly, D. Enhancing full-length antibody production by signal peptide engineering. *Microb. Cell Fact.* **2016**, *15*, 47. [CrossRef]
74. Saberianfar, R.; Chin-Fatt, A.; Scott, A.; Henry, K.A.; Topp, E.; Menassa, R. Plant-produced chimeric V_HH-sIgA against enterohemorrhagic *E. coli* intimin shows cross-serotype inhibition of bacterial adhesion to epithelial cells. *Front. Plant Sci.* **2019**, *10*, 270. [CrossRef] [PubMed]
75. Choi, J.H.; Lee, S.Y. Secretory and extracellular production of recombinant proteins using *Escherichia coli*. *Appl. Microbiol. Biotechnol.* **2004**, *64*, 625–635. [CrossRef] [PubMed]
76. Kinoshita, M.; Shimizu, M.; Akiyama, K.; Kato, H.; Moriyama, K.; Sawa, T. Epidemiological survey of serum titers from adults against various Gram-negative bacterial V-antigens. *PLoS ONE* **2020**, *15*, e0220924. [CrossRef] [PubMed]
77. Moody, A.M. The value of bacterial vaccines in immunization and therapy. *J. Am. Med. Assoc.* **1920**, *74*, 391–392. [CrossRef]
78. Meyer, K.F.; Cavanaugh, D.C.; Bartelloni, P.J.; Marshall, J.D. Plague immunization. I. Past and present trends. *J. Infect. Dis.* **1974**, *129*, S13–S18. [CrossRef] [PubMed]
79. Burrows, T.W.; Bacon, G.A. The effects of loss of different virulence determinants on the virulence and immunogenicity of strains of *Pasteurella pestis*. *Br. J. Exp. Pathol.* **1958**, *39*, 278–291. [PubMed]
80. Lawton, W.D.; Erdman, R.L.; Surgalla, M.J. Biosynthesis and purification of V and W antigen in *Pasteurella pestis*. *J. Immunol.* **1963**, *91*, 179–184.
81. Motin, V.L.; Nakajima, R.; Smirnov, G.B.; Brubaker, R.R. Passive immunity to yersiniae mediated by anti-recombinant V antigen and protein A-V antigen fusion peptide. *Infect. Immun.* **1994**, *62*, 4192–4201. [CrossRef]
82. Perry, R.D.; Harmon, P.A.; Bowmer, W.S.; Straley, S.C. A low-Ca2+ response operon encodes the V antigen of *Yersinia pestis*. *Infect. Immun.* **1986**, *54*, 428–434. [CrossRef]
83. Salmond, G.P.; Reeves, P.J. Membrane traffic wardens and protein secretion in Gram-negative bacteria. *Trends Biochem. Sci.* **1993**, *18*, 7–12. [CrossRef]
84. Cowan, C.; Philipovskiy, A.V.; Wulff-Strobel, C.R.; Ye, Z.; Straley, S.C. Anti-LcrV antibody inhibits delivery of Yops by *Yersinia pestis* KIM5 by directly promoting phagocytosis. *Infect. Immun.* **2005**, *73*, 6127–6137. [CrossRef] [PubMed]
85. Miller, N.C.; Quenee, L.E.; Elli, D.; Ciletti, N.A.; Schneewind, O. Polymorphisms in the LcrV gene of *Yersinia enterocolitica* and their effect on plague protective immunity. *Infect. Immun.* **2012**, *80*, 1572–1582. [CrossRef] [PubMed]
86. Ivanov, M.I.; Hill, J.; Bliska, J.B. Direct neutralization of type III effector translocation by the variable region of a monoclonal antibody to *Yersinia pestis* LcrV. *Clin. Vaccine Immun.* **2014**, *21*, 667–673. [CrossRef] [PubMed]
87. Xiao, X.; Zhu, Z.; Dankmeyer, J.L.; Wormald, M.M.; Fast, R.L.; Worsham, P.L.; Cote, C.K.; Amemiya, K.; Dimitrov, D.S. Human anti-plague monoclonal antibodies protect mice from *Yersinia pestis* in a bubonic plague model. *PLoS ONE* **2010**, *5*, e13047. [CrossRef] [PubMed]
88. Van Blarcom, T.J.; Sofer-Podesta, C.; Ang, J.; Boyer, J.L.; Crystal, R.G.; Georgiou, G. Affinity maturation of an anti-V antigen IgG expressed in situ through adenovirus gene delivery confers enhanced protection against *Yersinia pestis* challenge. *Gene Ther.* **2010**, *17*, 913–921. [CrossRef]
89. Zauberman, A.; Flashner, Y.; Levy, Y.; Vagima, Y.; Tidhar, A.; Cohen, O.; Bar-Haim, E.; Gur, D.; Aftalion, M.; Halperin, G.; et al. YopP-expressing variant of *Y. pestis* activates a potent innate immune response affording cross-protection against yersiniosis and tularemia. *PLoS ONE* **2013**, *8*, e83560. [CrossRef]
90. Philipovskiy, A.V.; Cowan, C.; Wulff-Strobel, C.R.; Burnett, S.H.; Kerschen, E.J.; Cohen, D.A.; Kaplan, A.M.; Straley, S.C. Antibody against V antigen prevents Yop-dependent growth of *Yersinia pestis*. *Infect. Immun.* **2005**, *73*, 1532–1542. [CrossRef]

91. Imamura, Y.; Yanagihara, K.; Fukuda, Y.; Kaneko, Y.; Seki, M.; Izumikawa, K.; Miyazaki, Y.; Hirakata, Y.; Sawa, T.; Wiener-Kronish, J.P.; et al. Effect of anti-PcrV antibody in a murine chronic airway *Pseudomonas aeruginosa* infection model. *Eur. Respir. J.* **2007**, *29*, 965–968. [CrossRef]
92. Song, Y.; Baer, M.; Srinivasan, R.; Lima, J.; Yarranton, G.; Bebbington, C.; Lynch, S.V. PcrV antibody-antibiotic combination improves survival in *Pseudomonas aeruginosa*-infected mice. *Eur. J. Clin. Microbiol. Infect. Dis.* **2012**, *31*, 1837–1845. [CrossRef]
93. Ranjbar, M.; Behrouz, B.; Norouzi, F.; Gargari, S.L.M. Anti-PcrV IgY antibodies protect against *Pseudomonas aeruginosa* infection in both acute pneumonia and burn wound models. *Mol. Immunol.* **2019**, *116*, 98–105. [CrossRef]
94. Frank, D.W.; Vallis, A.; Wiener-Kronish, J.P.; Roy-Burman, A.; Spack, E.G.; Mullaney, B.P.; Megdoud, M.; Marks, J.D.; Fritz, R.; Sawa, T. Generation and characterization of a protective monoclonal antibody to *Pseudomonas aeruginosa* PcrV. *J. Infect. Dis.* **2002**, *186*, 64–73. [CrossRef]
95. De Tavernier, E.; Detalle, L.; Morizzo, E.; Roobrouck, A.; De Taeye, S.; Rieger, M.; Verhaeghe, T.; Correia, A.; Van Hegelsom, R.; Figueirido, R.; et al. High throughput combinatorial formatting of PcrV nanobodies for efficient potency improvement. *J. Biol. Chem.* **2016**, *291*, 15243–15255. [CrossRef] [PubMed]
96. Jain, R.; Beckett, V.V.; Konstan, M.W.; Accurso, F.J.; Burns, J.L.; Mayer-Hamblett, N.; Milla, C.; VanDevanter, D.R.; Chmiel, J.F.; Chmiel, J.F.; et al. KB001-A, a novel anti-inflammatory, found to be safe and well-tolerated in cystic fibrosis patients infected with *Pseudomonas aeruginosa*. *J. Cyst. Fibros.* **2018**, *17*, 484–491. [CrossRef] [PubMed]
97. Warrener, P.; Varkey, R.; Bonnell, J.C.; DiGiandomenico, A.; Camara, M.; Cook, K.; Peng, L.; Zha, J.; Chowdury, P.; Sellman, B.; et al. A novel anti-PcrV antibody providing enhanced protection against *Pseudomonas aeruginosa* in multiple animal infection models. *Antimicrob. Agents Chemother.* **2014**, *58*, 4384–4391. [CrossRef] [PubMed]
98. Sawa, T.; Ito, E.; Nguyen, V.H.; Haight, M. Anti-PcrV antibody strategies against virulent *Pseudomonas aeruginosa*. *Hum. Vaccine Immunother.* **2014**, *10*, 2843–2852. [CrossRef]
99. Le, H.N.; Quetz, J.S.; Tran, V.G.; Le, V.T.M.; Aguiar-Alves, F.; Pinheiro, M.G.; Cheng, L.; Yu, L.; Sellman, B.R.; Stover, C.K.; et al. MEDI3902 correlates of protection against severe *Pseudomonas aeruginosa* pneumonia in a rabbit acute pneumonia model. *Antimicrob. Agents Chemother.* **2018**, *62*, e02565. [CrossRef]
100. Wang, Q.; Li, H.; Zhou, J.; Zhong, M.; Zhu, D.; Feng, N.; Liu, F.; Bai, C.; Song, Y. PcrV antibody protects multi-drug resistant *Pseudomonas aeruginosa* induced acute lung injury. *Respir. Physiol. Neurobiol.* **2014**, *193*, 21–28. [CrossRef]
101. Lynch, S.V.; Flanagan, J.L.; Sawa, T.; Fang, A.; Baek, M.S.; Rubio-Mills, A.; Ajayi, T.; Yanagihara, K.; Hirakata, Y.; Kohno, S.; et al. Polymorphisms in the *Pseudomonas aeruginosa* type III secretion protein, PcrV-Implications for anti-PcrV immunotherapy. *Microb. Pathog.* **2010**, *48*, 197–204. [CrossRef]
102. Warr, G.W.; Magor, K.E.; Higgins, D.A. IgY: Clues to the origins of modern antibodies. *Immunol. Today* **1995**, *16*, 392–398. [CrossRef]
103. Kinoshita, M.; Kato, H.; Yasumoto, H.; Shimizu, M.; Hamaoka, S.; Naito, Y.; Akiyama, K.; Moriyama, K.; Sawa, T. The prophylactic effects of human IgG derived from sera containing high anti-PcrV titers against pneumonia-causing *Pseudomonas aeruginosa*. *Hum. Vaccine Immunother.* **2016**, *12*, 2833–2846. [CrossRef]
104. François, B.; Luyt, C.E.; Dugard, A.; Wolff, M.; Diehl, J.L.; Jaber, S.; Forel, J.M.; Garot, D.; Kipnis, E.; Mebazaa, A.; et al. Safety and pharmacokinetics of an anti-PcrV PEGylated monoclonal antibody fragment in mechanically ventilated patients colonized with *Pseudomonas aeruginosa*: A randomized, double-blind, placebo-controlled trial. *Crit. Care Med.* **2012**, *40*, 2320–2326. [CrossRef] [PubMed]
105. Milla, C.E.; Chmiel, J.F.; Accurso, F.J.; Vandevanter, D.R.; Konstan, M.W.; Yarranton, G.; Geller, D.E. Anti-PcrV antibody in cystic fibrosis: A novel approach targeting *Pseudomonas aeruginosa* airway infection. *Pediatr. Pulmonol.* **2014**, *49*, 650–658. [CrossRef] [PubMed]
106. Tabor, D.E.; Oganesyan, V.; Keller, A.E.; Yu, L.; McLaughlin, R.E.; Song, E.; Warrener, P.; Rosenthal, K.; Esser, M.; Qi, Y.; et al. *Pseudomonas aeruginosa* PcrV and Psl, the molecular targets of bispecific antibody MEDI3902, are conserved among diverse global clinical isolates. *J. Infect. Dis.* **2018**, *218*, 1983–1994. [CrossRef] [PubMed]
107. Ali, S.O.; Yu, X.Q.; Robbie, G.J.; Wu, Y.; Shoemaker, K.; Yu, L.; DiGiandomenico, A.; Keller, A.E.; Anude, C.; Hernandez-Illas, M.; et al. Phase 1 study of MEDI3902, an investigational anti-*Pseudomonas aeruginosa* PcrV and Psl bispecific human monoclonal antibody, in healthy adults. *Clin. Microbiol. Infect.* **2019**, *25*, e1–e6. [CrossRef]

108. La Ragione, R.M.; Patel, S.; Maddison, B.; Woodward, M.J.; Best, A.; Whitelam, G.C.; Gough, K.C. Recombinant anti-EspA antibodies block *Escherichia coli* O157:H7-induced attaching and effacing lesions in vitro. *Microbes Infect.* **2006**, *8*, 426–433. [CrossRef]
109. Girard, F.; Batisson, I.; Martinez, G.; Breton, C.; Harel, J.J.; Fairbrother, J.M. Use of virulence factor-specific egg yolk-derived immunoglobulins as a promising alternative to antibiotics for prevention of attaching and effacing *Escherichia coli* infections. *FEMS Immunol. Med. Microbiol.* **2006**, *46*, 340–350. [CrossRef]
110. Cook, S.R.; Maiti, P.K.; DeVinney, R.; Allen-Vercoe, E.; Bach, S.J.; McAllister, T.A. Avian- and mammalian-derived antibodies against adherence-associated proteins inhibit host cell colonization by *Escherichia coli* O157:H7. *J. Appl. Microbiol.* **2007**, *103*, 1206–1219. [CrossRef]
111. Yu, S.; Gu, J.; Wang, H.; Wang, Q.; Luo, P.; Wu, C.; Zhang, W.; Guo, G.; Tong, W.; Zou, Q.; et al. Identification of a novel linear epitope on EspA from enterohemorrhagic *E. coli* using a neutralizing and protective monoclonal antibody. *Clin. Immunol.* **2010**, *138*, 77–84. [CrossRef]
112. Praekelt, U.; Reissbrodt, R.; Kresse, A.; Pavankumar, A.; Sankaran, K.; James, R.; Jesudason, M.; Anandan, S.; Prakasam, A.; Balaji, V.; et al. Monoclonal antibodies against all known variants of EspA: Development of a simple diagnostic test for enteropathogenic *Escherichia coli* based on a key virulence factor. *J. Med. Microbiol.* **2014**, *63*, 1595–1607. [CrossRef]
113. Desin, T.S.; Mickael, C.S.; Lam, P.K.; Potter, A.A.; Köster, W. Protection of epithelial cells from *Salmonella enterica* serovar Enteritidis invasion by antibodies against the SPI-1 type III secretion system. *Can. J. Microbiol.* **2010**, *56*, 522–526. [CrossRef]
114. Barta, M.L.; Guragain, M.; Adam, P.; Dickenson, N.E.; Patil, M.; Geisbrecht, B.V.; Picking, W.L.; Picking, W.D. Identification of the bile salt binding site on IpaD from *Shigella flexneri* and the influence of ligand binding on IpaD structure. *Proteins* **2012**, *80*, 935–945. [CrossRef] [PubMed]
115. Barta, M.L.; Shearer, J.P.; Arizmendi, O.; Tremblay, J.M.; Mehzabeen, N.; Zheng, Q.; Battaile, K.P.; Lovell, S.; Tzipori, S.; Picking, W.D.; et al. Single-domain antibodies pinpoint potential targets within *Shigella* invasion plasmid antigen D of the needle tip complex for inhibition of type III secretion. *J. Biol. Chem.* **2017**, *292*, 16677–16687. [CrossRef]
116. Ivanov, M.I.; Noel, B.L.; Rampersaud, R.; Mena, P.; Benach, J.L.; Bliska, J.B. Vaccination of mice with a Yop translocon complex elicits antibodies that are protective against infection with $F1^{-}$ *Yersinia pestis*. *Infect. Immun.* **2008**, *76*, 5181–5190. [CrossRef]
117. Guirro, M.; de Souza, R.L.; Piazza, R.M.F.; Guth, B.E.C. Antibodies to intimin and *Escherichia coli*-secreted proteins EspA and EspB in sera of Brazilian children with hemolytic uremic syndrome and healthy controls. *Vet. Immunol. Immunopathol.* **2013**, *152*, 121–125. [CrossRef] [PubMed]
118. Rabinovitz, B.C.; Larzábal, M.; Vilte, D.A.; Cataldi, A.; Mercado, E.C. The intranasal vaccination of pregnant dams with Intimin and EspB confers protection in neonatal mice from *Escherichia coli* (EHEC) O157: H7 infection. *Vaccine* **2016**, *34*, 2793–2797. [CrossRef] [PubMed]
119. Forbes, S.J.; Eschmann, M.; Mantis, N.J. Inhibition of *Salmonella enterica* serovar Typhimurium motility and entry into epithelial cells by a protective antilipopolysaccharide monoclonal immunoglobulin A antibody. *Infect. Immun.* **2008**, *76*, 4137–4144. [CrossRef]
120. Forbes, S.J.; Martinelli, D.; Hsieh, C.; Ault, J.G.; Marko, M.; Mannella, C.A.; Mantis, N.J. Association of a protective monoclonal IgA with the O antigen of *Salmonella enterica* serovar Typhimurium impacts type 3 secretion and outer membrane integrity. *Infect. Immun.* **2012**, *80*, 2454–2463. [CrossRef]
121. Goodin, J.L.; Raab, R.W.; McKown, R.L.; Coffman, G.L.; Powell, B.S.; Enama, J.T.; Ligon, J.A.; Andrews, G.P. *Yersinia pestis* outer membrane type III secretion protein YscC: Expression, purification, characterization, and induction of specific antiserum. *Protein Expr. Purif.* **2005**, *40*, 152–163. [CrossRef]
122. Costa, T.R.; Felisberto-Rodrigues, C.; Meir, A.; Prevost, M.S.; Redzej, A.; Trokter, M.; Waksman, G. Secretion systems in Gram-negative bacteria: Structural and mechanistic insights. *Nat. Rev. Microbiol.* **2015**, *13*, 343–359. [CrossRef]
123. Arnoldo, A.; Curak, J.; Kittanakom, S.; Chevelev, I.; Lee, V.T.; Sahebol-Amri, M.; Koscik, B.; Ljuma, L.; Roy, P.J.; Bedalov, A.; et al. Identification of small molecule inhibitors of *Pseudomonas aeruginosa* exoenzyme S using a yeast phenotypic screen. *PLoS Genet.* **2008**, *4*, e1000005. [CrossRef]
124. Akopyan, K.; Edgren, T.; Wang-Edgren, H.; Rosqvist, R.; Fahlgren, A.; Wolf-Watz, H. Translocation of surface-localized effectors in type III secretion. *Proc. Natl. Acad. Sci. USA* **2011**, *108*, 1639–1644. [CrossRef] [PubMed]

125. Singh, A.K.; Kingston, J.J.; Murali, H.S.; Batra, H.V. A recombinant bivalent fusion protein rVE confers active and passive protection against *Yersinia enterocolitica* infection in mice. *Vaccine* **2014**, *32*, 1233–1239. [CrossRef] [PubMed]
126. González-Juarbe, N.; Shen, H.; Bergman, M.A.; Orihuela, C.J.; Dube, P.H. YopE specific CD8+ T cells provide protection against systemic and mucosal *Yersinia pseudotuberculosis* infection. *PLoS ONE* **2017**, *12*, e0172314. [CrossRef] [PubMed]
127. Kerschen, E.J.; Cohen, D.A.; Kaplan, A.M.; Straley, S.C. The plague virulence protein YopM targets the innate immune response by causing a global depletion of NK cells. *Infect. Immun.* **2004**, *72*, 4589–4602. [CrossRef] [PubMed]
128. Rüter, C.; Silva, M.R.; Grabowski, B.; Lubos, M.L.; Scharnert, J.; Poceva, M.; von Tils, D.; Flieger, A.; Heesemann, J.; Bliska, J.B.; et al. Rabbit monoclonal antibodies directed at the T3SS effector protein YopM identify human pathogenic *Yersinia* isolates. *Int. J. Med. Microbiol.* **2014**, *304*, 444–451. [CrossRef] [PubMed]
129. Pacheco, A.R.; Lazarus, J.E.; Sit, B.; Schmieder, S.; Lencer, W.I.; Blondel, C.J.; Doench, J.G.; Davis, B.M.; Waldor, M.K. CRISPR screen reveals that EHEC's T3SS and Shiga toxin rely on shared host factors for infection. *MBio* **2018**, *9*, e01003-18. [CrossRef]
130. Ruano-Gallego, D.; Yara, D.A.; Di Ianni, L.; Frankel, G.; Schüller, S.; Fernández, L.Á. A nanobody targeting the translocated intimin receptor inhibits the attachment of enterohemorrhagic *E. coli* to human colonic mucosa. *PLoS Pathog.* **2019**, *15*, e1008031. [CrossRef]
131. Kühne, S.A.; Hawes, W.S.; La Ragione, R.M.; Woodward, M.J.; Whitelam, G.C.; Gough, K.C. Isolation of recombinant antibodies against EspA and intimin of *Escherichia coli* O157:H7. *J. Clin. Microbiol.* **2004**, *42*, 2966–2976. [CrossRef]
132. Jones-Carson, J.; McCollister, B.D.; Clambey, E.T.; Vázquez-Torres, A. Systemic CD8 T-cell memory response to a *Salmonella* pathogenicity island 2 effector is restricted to *Salmonella enterica* encountered in the gastrointestinal mucosa. *Infect. Immun.* **2007**, *75*, 2708–2716. [CrossRef]
133. Turbyfill, K.R.; Hartman, A.B.; Oaks, E.V. Isolation and characterization of a *Shigella flexneri* invasion complex subunit vaccine. *Infect. Immun.* **2000**, *68*, 6624–6632. [CrossRef]
134. Desin, T.S.; Townsend, H.G.; Potter, A.A. Antibodies directed against Shiga-toxin producing *Escherichia coli* serotype O103 type III secreted proteins block adherence of heterologous STEC serotypes to HEp-2 cells. *PLoS ONE* **2015**, *10*, e0139803. [CrossRef]
135. Engel, J.; Balachandran, P. Role of *Pseudomonas aeruginosa* type III effectors in disease. *Curr. Opin. Microbiol.* **2009**, *12*, 61–66. [CrossRef]
136. Corech, R.; Rao, A.; Laxova, A.; Moss, J.; Rock, M.J.; Li, Z.; Kosorok, M.R.; Splaingard, M.L.; Farrell, P.M.; Barbieri, J.T. Early immune response to the components of the type III system *of Pseudomonas aeruginosa* in children with cystic fibrosis. *J. Clin. Microbiol.* **2005**, *43*, 3956–3962. [CrossRef] [PubMed]
137. Finco, O.; Frigimelica, E.; Buricchi, F.; Petracca, R.; Galli, G.; Faenzi, E.; Meoni, E.; Bonci, A.; Agnusdei, M.; Nardelli, F.; et al. Approach to discover T- and B-cell antigens of intracellular pathogens applied to the design of *Chlamydia trachomatis* vaccines. *Proc. Natl. Acad. Sci. USA* **2011**, *108*, 9969–9974. [CrossRef] [PubMed]
138. Tsai, P.Y.; Hsu, M.C.; Huang, C.T.; Li, S.Y. Human antibody and antigen response to IncA antibody of *Chlamydia trachomatis. Int. J. Immunopathol. Pharmacol.* **2007**, *20*, 155–161. [CrossRef]
139. Freeman, J.A.; Rappl, C.; Kuhle, V.; Hensel, M.; Miller, S.I. SpiC is required for translocation of *Salmonella* pathogenicity island 2 effectors and secretion of translocon proteins SseB and SseC. *J. Bacteriol.* **2002**, *184*, 4971–4980. [CrossRef]
140. Geng, S.; Qian, S.; Pan, Z.; Sun, L.; Chen, X.; Jiao, X. Preparation of monoclonal antibodies against SpiC protein secreted by T3SS-2 of *Salmonella* spp. *Monoclon. Antib. Immunodiagn. Immunother.* **2015**, *34*, 432–435. [CrossRef]
141. Singh, K.; Ejaz, W.; Dutta, K.; Thayumanavan, S. Antibody delivery for intracellular targets: Emergent therapeutic potential. *Bioconjugate Chem.* **2019**, *30*, 1028–1041. [CrossRef]
142. Briers, Y.; Walmagh, M.; Van Puyenbroeck, V.; Cornelissen, A.; Cenens, W.; Aertsen, A.; Oliveira, H.; Azeredo, J.; Verween, G.; Pirnay, J.P.; et al. Engineered endolysin-based "Artilysins" to combat multidrug-resistant Gram-negative pathogens. *MBio* **2014**, *5*, e01379. [CrossRef]
143. Alzogaray, V.; Danquah, W.; Aguirre, A.; Urrutia, M.; Berguer, P.; Véscovi, E.G.; Haag, F.; Koch-Nolte, F.; Goldbaum, F.A. Single-domain llama antibodies as specific intracellular inhibitors of SpvB, the actin ADP-ribosylating toxin of *Salmonella* Typhimurium. *FASEB J.* **2011**, *25*, 526–534. [CrossRef]

144. Winstanley, C.; Hart, C.A. Type III secretion systems and pathogenicity islands. *J. Med. Microbiol.* **2001**, *50*, 116–126. [CrossRef] [PubMed]
145. Zhang, Y.; Sun, X.; Qian, Y.; Yi, H.; Song, K.; Zhu, H.; Zonta, F.; Chen, W.; Ji, Q.; Miersch, S.; et al. A potent Anti-SpuE antibody allosterically inhibits type III secretion system and attenuates virulence of *Pseudomonas aeruginosa*. *J. Mol. Biol.* **2019**, *431*, 4882–4896. [CrossRef] [PubMed]
146. Wang, J.; Wang, J.; Zhang, L.H. Immunological blocking of spermidine-mediated host–pathogen communication provides effective control against *Pseudomonas aeruginosa* infection. *Microb. Biotechnol.* **2020**, *13*, 87–96. [CrossRef] [PubMed]

Review

Principles of *N*-Linked Glycosylation Variations of IgG-Based Therapeutics: Pharmacokinetic and Functional Considerations

Souad Boune, Peisheng Hu, Alan L. Epstein and Leslie A. Khawli *

Department of Pathology, Keck School of Medicine, University of Southern California, Los Angeles, CA 90089, USA; so3ad86@gmail.com (S.B.); peisheng@usc.edu (P.H.); aepstein@usc.edu (A.L.E.)

* Correspondence: lkhawli@usc.edu

Received: 13 May 2020; Accepted: 3 June 2020; Published: 10 June 2020

Abstract: The development of recombinant therapeutic proteins has been a major revolution in modern medicine. Therapeutic-based monoclonal antibodies (mAbs) are growing rapidly, providing a potential class of human pharmaceuticals that can improve the management of cancer, autoimmune diseases, and other conditions. Most mAbs are typically of the immunoglobulin G (IgG) subclass, and they are glycosylated at the conserved asparagine position 297 (Asn-297) in the CH2 domain of the Fc region. Post-translational modifications here account for the observed high heterogeneity of glycoforms that may or not impact the stability, pharmacokinetics (PK), efficacy, and immunogenicity of mAbs. These modifications are also critical for the Fc receptor binding, and consequently, key antibody effector functions including antibody-dependent cell-mediated cytotoxicity (ADCC) and complement-dependent cytotoxicity (CDC). Moreover, mAbs produced in non-human cells express oligosaccharides that are not normally found in serum IgGs might lead to immunogenicity issues when administered to patients. This review summarizes our understanding of the terminal sugar residues, such as mannose, sialic acids, fucose, or galactose, which influence therapeutic mAbs either positively or negatively in this regard. This review also discusses mannosylation, which has significant undesirable effects on the PK of glycoproteins, causing a decreased mAbs' half-life. Moreover, terminal galactose residues can enhance CDC activities and Fc–C1q interactions, and core fucose can decrease ADCC and Fc–FcγRs binding. To optimize the therapeutic use of mAbs, glycoengineering strategies are used to reduce glyco-heterogeneity of mAbs, increase their safety profile, and improve the therapeutic efficacy of these important reagents.

Keywords: glycosylation; post-translational modifications; pharmacokinetics; effector functions; antibody-dependent cell-mediated cytotoxicity; complement-dependent cytotoxicity; immunogenicity; pharmacodynamics; glycoengineering; antibody-drug conjugates

1. Introduction

Monoclonal antibody (mAb)-based therapeutics have been increasingly studied and utilized as therapeutic agents for the past 20 years [1]. Even though mAb technology was invented early in 1975 by Milstein and Koehler [2], the potential of these agents was not appreciated originally because of anti-drug antibody (ADA) responses in humans induced by murine antibodies [3]. However, with the rapid growth of biotechnology-derived techniques and the advanced knowledge of the immune system, scientists have realized the roll that mAbs can play in the treatment of many diseases [4]. Today, there are more than 60 products of therapeutic monoclonal antibodies (mAbs) that are approved in the US for human use, about 240 in clinical testing, and around 40 entering clinical trials each year [5,6].

Therapeutic antibodies are generally IgGs. An IgG is a glycoprotein that contains four polypeptide chains: Two identical heavy chains (H) and two identical light chains (L). The light and heavy chains pair by covalent disulfide bonds and noncovalent associations (Figure 1) [4]. Each heavy chain is connected to one light chain by one disulfide bond. Each antibody molecule is made of three globular domain structures forming a "Y" shape, two of which are the fragments that bind to the antigens (Fab) and the other is the fragment crystallizable (Fc) for the activation of Fcγ receptors (FcγRs) on leukocytes and the C1 component of complement [6]. IgG molecules bear *N*-glycosylation at the conserved asparagine at position 297 (Asn-297) in the heavy chain of the CH2 constant domain of the Fc region [6]. The oligosaccharide is an essential player in Fc effector functions including antibody-dependent cellular cytotoxicity (ADCC) and complement-dependent cytotoxicity (CDC), which are major mechanisms of action of therapeutic antibodies located in the Fc region. Alteration of glycan compositions and structures can impact the effector function by causing conformational changes of the Fc domain, which would affect binding affinity to Fcγ receptors [3,5]. Thus, engineering of Fc glycosylation to develop therapeutic monoclonal antibodies with desired characteristics is a promising strategy to enhance functionality and efficacy of therapeutic IgG antibodies. In this review, Fc *N*-glycan structure and biosynthesis are briefly reviewed, followed by a discussion of the knowledge acquired recently about the influence of glycosylation of antibodies on therapeutic antibody immunogenicity, pharmacokinetics (PK), and effector functions. Furthermore, current Fc glycoengineering strategies used to produce mAbs with higher homogeneity and effector functions are introduced and discussed. In the following sections we will also discuss those aspects of glycosylation variations which relate to the PK and pharmacodynamic (PD) parameters of currently approved antibody-based therapeutics.

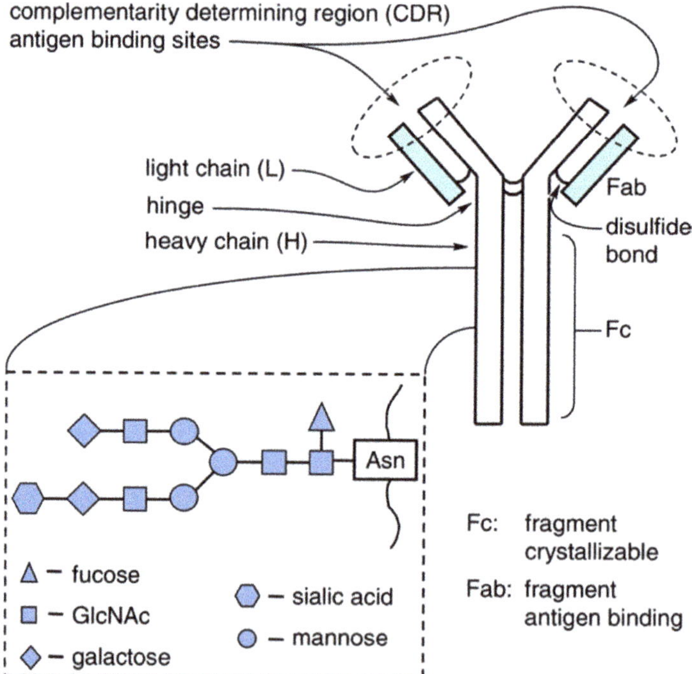

Figure 1. Simplified structure of an immunoglobulin (IgG). Inset shows an example of an IgG Fc diantennary oligosaccharide, which in normal IgG, is attached at an asparagine residue at position 297 (Asn-297). Generally, the oligosaccharide has a core pentasaccharide with varying addition of galactose, fucose, sialic acid, and *N*-acetylglucosamine (GlcNAc). Reproduced from Bakhtiar, 2012 [4] with permission of the copyright owner.

2. IgG Glycan Structure and Biosynthesis

Post-translational modification is a biological process that involves the modification of an amino acid side chain, terminal amino, or carboxyl group by means of covalent or enzymatic modifications following IgG biosynthesis. Generally, these modifications may include phosphorylation, acetylation, glycosylation, sialylation of one or more amino acids in the protein, and also may include the formation of S-S bridges between 2 SH groups on amino acids, and proteolysis. Post-translational modifications contribute to the final tertiary (three-dimensional) structure of IgGs and play a key role in the biological activity and interaction with other cellular molecules such as proteins, nucleic acids, lipids, and cofactors. These modifications are not predictable by the sequence of IgG and are often critical in determining the way IgG behaves (e.g., its function and degradation). Therefore, each therapeutic protein will have a unique post-translational modification profile in its natural state, and as discussed further in this review, the post-translational modification profile of an IgG can potentially impact drug stability, safety, and efficacy.

2.1. IgG Glycan Structure

Structurally, the *N*-linked glycans of human IgGs are typically biantennary complexes. Different residues, such as fucose, bisecting GlcNAc, galactose, and sialic acid, can be added to this core biantennary complex structure (GlcNAc2Man3GlcNAc2), generating heterogeneity of the IgG-Fc glycans of normal polyclonal IgGs [5,7]. The heterogeneous glycans can be classified into three sets (G0, G1, and G2), depending on the number of galactose residues in the outer arms of biantennary glycans. Within each of these sets, there are different species that arise from the presence or absence of core fucose and bisecting GlcNAc (Figure 2) [3].

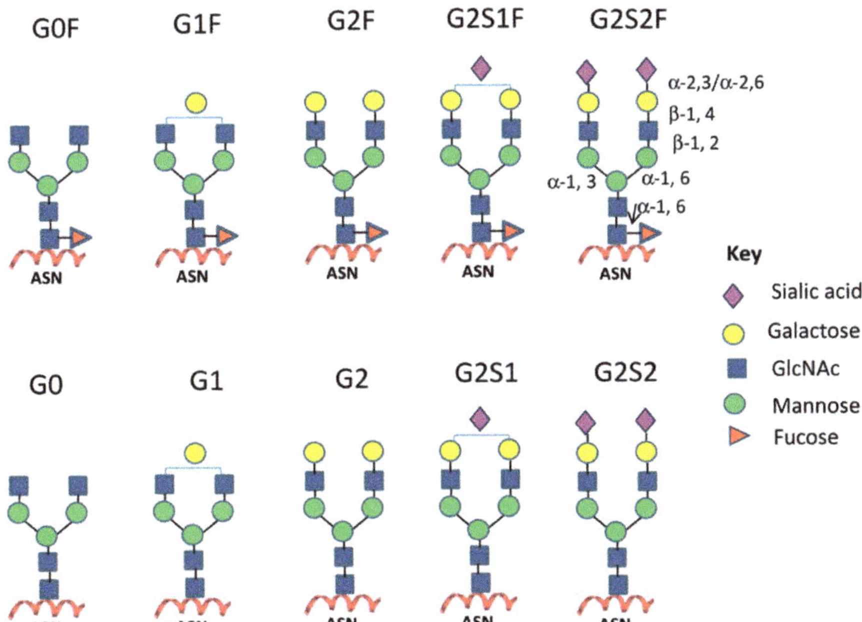

Figure 2. Major *N*-linked glycoforms of therapeutic monoclonal antibodies (mAbs). Reproduced from Liu, 2015 [3] with permission of the copyright owner.

2.2. Glycan Biosynthesis in Human Cells

Glycosylation is the most common post-translational modification of proteins. It is a complex process that results in a great diversity of carbohydrate–protein bonds and glycan structures. It is known that it has a great impact on protein structures and functions [8]. Glycosylation of IgG is an enzyme-directed chemical reaction that occurs in the endoplasmic reticulum (ER) and the Golgi apparatus of the cell. Initially, a Glc3Man9GlcNAc2 oligosaccharide is transferred to Asn-297 of the IgG heavy chain via an oligosaccharyltransferase complex in the ER. Subsequently, the *N*-glycans are subjected to a sequence of consecutive modifications by sets of glycosidases and glycosyltransferases [9]. Polypeptide-associated Glc3Man9GlcNAc2 is trimmed by glucosidases I and II and endo-mannosidase in the lumen of the ER, resulting in the removal of three Glc residues and a mannose residue to produce Man8GlcNAc2 (Figure 3) [5]. In the cis-Golgi, the Man8GlcNAc2 is sequentially subjected to two class I α-mannosidases that act particularly on α-1,2-Man residues to produce the core Man5GlcNAc2 glycan for additional modification in the medial and trans-Golgi, mediated by GlcNAc transferases I, II, and III (GnT I, II, and III), α-1,6-fucosyltransferase (FUT8), galactosyltransferases (GalT), and sialyltransferases (SiaT) [3,5,9].

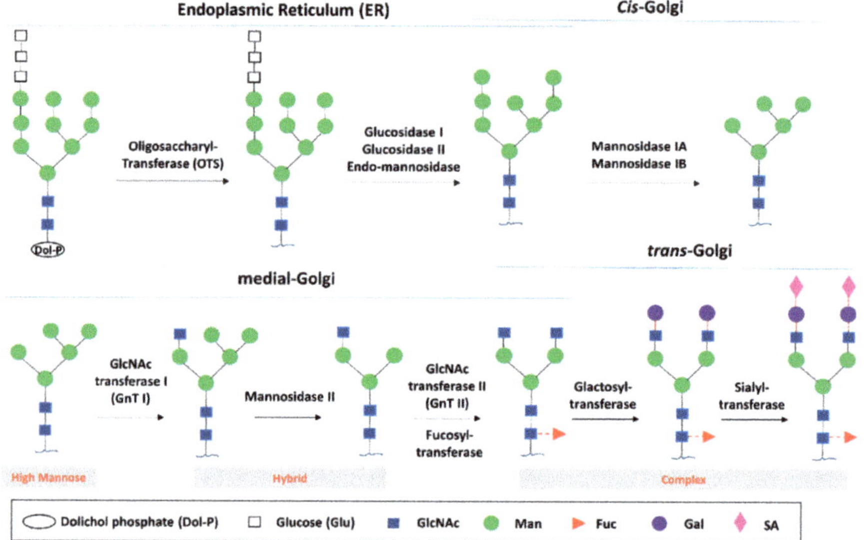

Figure 3. Glycan biosynthesis through the endoplasmic reticulum (ER) and Golgi glycosylation pathways. The biosynthesis begins with the processing of the initial high mannose *N*-glycan in the ER, followed by transferring into the cis-Golgi to generate the core *N*-glycan substrate used for further diversification in the trans-Golgi. The potential glycoforms include the high mannose, hybrid, and complex structure. Reprinted from Li et al., 2017 [5] with permission of the copyright owner.

3. *N*-Glycosylation Impact on mAb Structure and Effector Function

The amount and nature of glycosylation can dramatically affect the behavior of endogenous and recombinant IgGs. The most commonly described roles for glycosylation are related to receptor binding and Fc effector functions. However, the glycosylation profile of an IgG can also substantially affect its PK and distribution. In order to understand the possible manipulations and reasons behind glycosylation and glycoengineering, the reader is also directed to references [3–7,10] for a thorough overview describing the current understanding of glycosylation pattern (and normal variation), normal PK, and effector functions in IgG. As such, Fc glycosylation has great influence on mAbs' efficacy, stability, safety, immunogenicity, PK, and PD.

3.1. Impact of Fc Glycosylation on Structure

It is well established that the glycan structures can directly affect IgG through altering the conformation of the Fc domain [11]. *N*-glycans have essential structural supportive functions. They play a critical role in the stability of CH2 domain of IgGs, which binds to the glycans via extensive non-covalent interactions that reduce the dynamics of CH2 and aid in CH2 folding. Deglycosylation makes mAbs thermally less stable and more prone to unfolding and degradation [10]. Furthermore, removal of sugar residues leads to the generation of a "closed" conformation while the fully galactosylated IgG-Fc correlates with "open" conformation, which may be most favorable for FcγR binding [12] (Figure 4) [6].

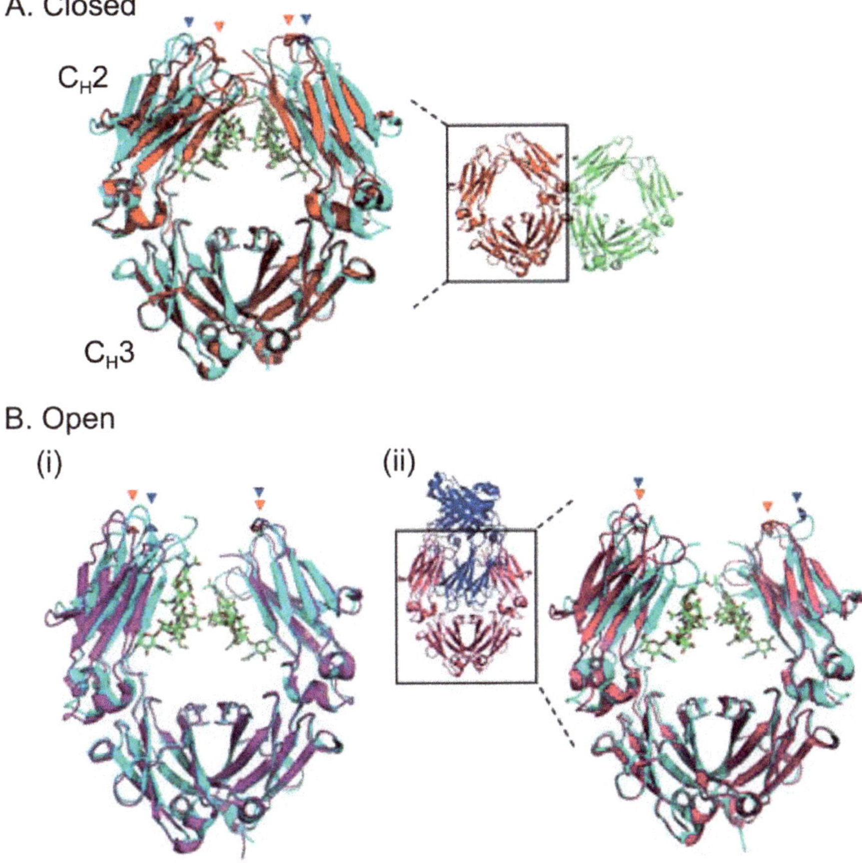

Figure 4. Comparison of non-glycosylated and glycosylated Fc structures. (**A**) Closed conformation of the non-glycosylated Fc. Overall structure of the two aglycosylated Fc molecules is shown in red and green, and the Fc shown in red is superimposed with the glycosylated Fc. (**B**) Open conformation of the non-glycosylated Fc. Overall structure of the two interlocked Fc molecules is shown in pink and blue. The Fc shown in pink is superimposed with the glycosylated Fc. The Fc glycans are shown in green sticks. The Pro329 residues located in the FG loop of the CH2 domains are indicated by red and blue arrowheads for the non-glycosylated and glycosylated CH2 domains, respectively. Reproduced from Mimura et al., 2018 [6] with permission of the copyright owner.

3.2. Impact of Fc Glycosylation on Immunogenicity

As mentioned above, glycosylated mAbs can alter their safety and immunogenicity. Glycan patterns are highly variable since they depend on the host glycosylation machinery. Thus, different host cells can produce different recombinant antibodies with different glycoforms. Most therapeutic recombinant antibodies are Chinese hamster ovary (CHO)-derived recombinant IgG molecules, and some are made in murine myeloma cell lines NS0 and SP2/0. Recombinant antibodies produced in CHO cells are glycosylated similarly to natural human IgG. On the other hand, recombinant human IgGs derived in murine myeloma cells can have different glycoforms because they add sugars which are not normally found in the human IgG [3,5]. Terminal-sugar residues expressed in non-human glycoforms that are not normally found in endogenous serum IgGs could be highly immunogenic in humans [13]. Immunogenicity of these therapeutic antibodies can lead to reduced efficacy and safety and cause anti-drug Ab responses (ADA) and hypersensitivity reactions. Therefore, the expression system (bacteria, yeast, insect, plant, or mammalian cells) that is used to generate recombinant mAbs is crucial and has tremendous influence on the mAb function in vivo [14].

Glycoproteins that are produced in yeasts, plants, and insect cells usually have high-mannose contents, which can increase immunogenicity of recombinant mAbs [15]. Lam et al. have demonstrated that antigen mannosylation significantly increases protein immunogenicity in mice [16]. Most therapeutic mAbs, however, have very low levels of high-mannose content [17]. Moreover, terminal sialic acids of therapeutic mAbs derived in non-human cells, such as murine myeloma cell lines, have been shown to be a possible factor that cause immunogenicity in patients since they express the N-glycolylneuraminic acid (NGNA) form of sialic acids that are not normally found in human IgGs [18]. The main reason behind this significant immunogenicity could be NGNA-specific antibodies that have been found to be expressed by all humans [19]. More specific investigations by Qian and coworkers have reported that Cetuximab, a murine myeloma cell-derived novel therapeutic monoclonal antibody that contains NGNA, caused immune interaction with NGNA-specific antibodies [20]. Because of these findings, assessment of the immunogenicity of therapeutic Abs is a critical quality attribute that should be considered with respect to the manufacturing of these therapeutic glycoproteins.

3.3. Impact of Fc Glycosylation on Pharmacokinetics

Clearance has a critical impact on the efficacy of therapeutic antibodies. Monoclonal antibodies are high-molecular weight drugs that are large complex proteins (approximately 150 kDa) that are not eliminated through kidney filtration. In addition, they can escape fast degradation in the lysosomes through the neonatal Fc receptor (FcRn) recycling mechanism [21]. The binding of Fc to the neonatal Fc receptor at the CH2–CH3 domain plays a critical role in the PK properties of IgG molecules. Recycling of antibodies results in long half-life of IgGs in the serum (up to 4 weeks) [22]. Roopenian et al. conducted experiments on FcRn knockout mice and they have concluded that FcRn is responsible for protecting IgG from catabolism [22]. Both glycosylated and deglycosylated IgGs bind equally to the (FcRn) receptor [23]. Therefore, the interaction between FcRn and IgG is independent of the Fc glycans due to their protected and buried position within the antibody structure. The significance of Fc glycosylation in the PK of therapeutic mAbs can be examined by comparing the biological activities of glycosylated IgG with either enzymatically deglycosylated IgGs or by preparing aglycosylated IgGs (bearing Asn-297 mutation) using molecular biology techniques. Several studies have compared the biological activity and PK properties of antibodies with different glycoforms in humans and animals [24,25].

Liu et al. confirmed that glycosylation is not required for an IgG antibody's long half-life after they characterized aglycosylated IgGs by chemical modification and genetic engineering [23]. These animal studies demonstrated that the PK profile of an aglycosylated IgG1 mAb with an Asn-297 mutation was almost identical to that of the glycosylated form. Another clinical trial conducted in 2009 by Clarke et al. also demonstrated that aglycosylated mAb ALD518 (clazakizumab), a humanized anti-human

IL-6 IgG1 produced in yeast, had a normal PK in humans and animals. In their phase I clinical trial, the circulating half-life for ALD518 was 20–32 days, which is consistent with the half-life of a normal human IgG1 [26]. Moreover, Abuqayyas and colleagues found that 8C2, a mouse IgG mAb, exhibited similar PK and tissue distribution in both FcγR knockout mice and in wild type mice [27]. Similar PK properties of glycosylated and non-glycosylated IgGs confirm that antibody clearance in humans and animals is not significantly affected by Fc glycan removal [10,24,28].

3.4. Effect of Terminal Mannose on Pharmacokinetics

Circulating glycoproteins can be cleared from the blood by receptors that recognize specific glycan forms. Glycan receptors that are involved in the clearance of glycoproteins include the mannose receptor (ManR) and the asialoglycoprotein receptor (ASGPR). The asialoglycoprotein receptors bind to terminal Gal residues and the ManR bind to glycoproteins with terminal Man or GlcNac sugars. Glycan binding to these receptors expressed on tissues was considered to have potential effects on the PK of antibodies bearing these terminal sugars and to cause faster removal from circulation [25]. Consistent with this, Kanda et al. demonstrated that IgG antibodies with high-mannose glycoforms have shorter half-life compared to those with the complex-type glycans in mice [29]. Yu et al. conducted a PK study in mice, and they determined the clearance rate of antibodies bearing Man8/9 and Man5 glycan. They showed that the antibodies bearing the high mannose glycoform were cleared faster compared with antibodies bearing the fucosylated complex glycoform, while the PK properties of antibodies with Man8/9 and Man5 glycoforms appeared similar (Figure 5) [25]. In agreement with previous human studies, Goetze and coworkers observed faster elimination of therapeutic IgGs containing Fc high-mannose glycans from circulation compared to other glycoforms [17]. In addition, differences in high-mannose structural isoform clearance rates in humans were reported by Chen et al., but these investigators suggested that changes in the serum half-life of mAbs bearing high mannose glycoforms were actually due to glycan cleavage [24]. Another investigation done by Millward et al. reached contradictory conclusions. They found no significant difference in serum half-life in mice between high-mannose IgG type and complex IgGs [30]. In summary, high-terminal mannose content appears to be an important point that should be considered as it may affect PK properties and efficacy of therapeutic antibodies. Because of the above findings, most mAbs for clinical use possess relatively low-terminal high-mannose glycan content.

In general, glycans that have a major impact on PK of mAbs include mannose, sialic acids, galactose, and fucose [3,25] (Figure 5). The negatively-charged sialic acids attached to the terminus of glycan chains have been shown to affect half-life for many glycoproteins. It was found that IgGs with exposed terminal Gal (after removal of sialic acid) resulted in a decreased half-life in mice and localization in the liver [3]. To date, the PK properties of different glycan compositions in approved antibody-based therapeutics have not yet been investigated in the clinic.

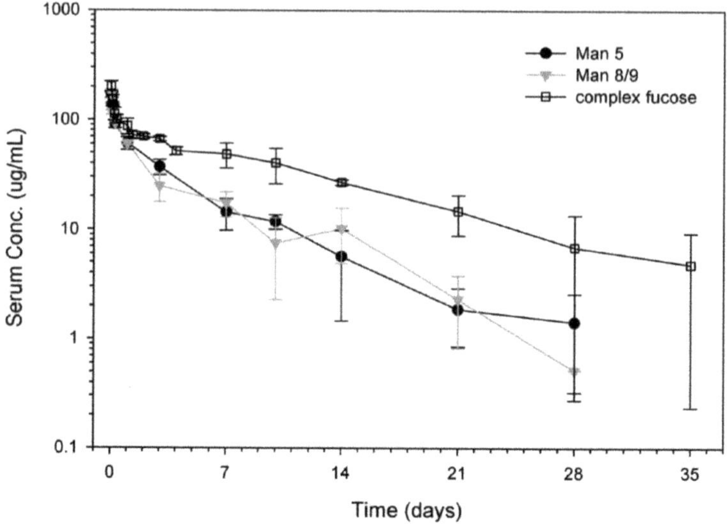

Figure 5. The influence of high-mannose glycans on the pharmacokinetics (PKs) of mAbs. Pharmacokinetic profiles of mAb variants in athymic nude mice. The two groups conducted in this study were mAb with >99% Man5 (solid black circle) and mAb with >99% Man8/9 (solid gray triangle). The complex-fucosylated profile (open squares) was from a separate study conducted in a similar fashion to the current study. Reproduced from Yu et al., 2012 [25] with permission of the copyright owner.

4. Impact of Fc Glycosylation on Pharmacodynamics

The oligosaccharides of the IgG-Fc play a critical role in activation of FcγRs and complement C1. FcγR-mediated effector functions result in the killing of the target cell. FcγRs are responsible for ADCC effector function and, while the receptor C1q mediates CDC. Many studies have found that the lack of glycosylation noticeably decreases the binding affinity to FcγRI and eliminates the binding to FcγRII and FcγRIII receptors [31,32].

4.1. Sialic Acid

Sialic acids are present in human serum IgGs as *N*-acetylneuraminic acid (NANA) attached to a terminal galactose by an α-2,3 or α-2,6 linkage. Recombinant monoclonal antibodies expressed in CHO cell line also have NANA, but it is only attached by α-2,3 linkage [18]. On the other hand, monoclonal antibodies produced in NS0 and SP2/0 cell lines have NGNA, a sialic acid form produced by hydroxylation of NANA utilizing cytidine monophosphate *N*-acetylneuraminic acid hydroxylase enzyme which is absent in human and CHO cells under normal conditions [33]. Typically, the level of sialic acid in human endogenous IgGs is ~11%–15% [18,34]. Studies to date that explore the effects of sialic acid on Fcγ receptors binding are inconclusive. Scallon and coworkers studied pairs of monoclonal human IgG Abs produced in mouse hybridoma cell lines with different amounts of sialic acid in their Fc glycans [35]. They demonstrated that a higher content of terminal sialylation was correlated with decreased activity in ADCC and lower-affinity binding to FcγRIIIa on natural killer (NK) cells in vitro. Similarly, Kaneko et al. reported that Fc sialylation affects antibody effector functions including reduction of ADCC in both in vitro and in vivo [36].

However, another in vitro study investigated the influence of sialic acid on IgG1 effector functions using different glycosylated forms of a single drug with various levels of sialylation generated by in vitro glycoengineering [37]. They found that terminal sialylation had no impact, neither positive nor negative, on ADCC activity, FcγRI, and RIIIa receptors, but slightly improved affinity to FcγRIIa was reported [37]. Furthermore, full sialylation of human monoclonal IgG1 was reported to interfere with the induction of CDC in vitro [38].

Recently, Fc sialylation has drawn scientists' attention as it has been attributed to increased anti-inflammatory responses to intravenous Ig (IVIG) for the treatment of autoimmune and inflammatory diseases [36,39]. IVIG suppresses inflammation by binding to inhibitory FcγRIIb. Sialylated IgG initiates anti-inflammatory effects by binding to the murine C-type lectin-like receptor-specific intracellular adhesion molecule-grabbing non-integrin R1 (SIGN-RI) (DCSIGN in humans) expressed by macrophage and dendritic cells. As a result, FcγRIIb expression will be upregulated and Treg cell populations will expand, leading to significant suppression of inflammatory responses [40,41]. Kaneko et al. approved that sialylated human IgG has elevated anti-inflammatory activity compared to the desialylated IgG utilizing a mouse model of rheumatoid arthritis [36]. However, these findings in mice were contradicted by a study of rheumatoid arthritis during pregnancy [42]. This study showed that remission of rheumatoid arthritis was associated with galactosylation independently of sialylation. In summary, sialylated glycans collectively have both positive and negative influences on IgG effector functions, making it crucial to quantitate the sialylation of mAbs headed for the clinic, especially to treat autoimmune conditions. To date, the functions of different Fc-sialylated glycans in approved antibody-based therapeutics have not yet been investigated in the clinic.

4.2. Terminal Galactose

Recombinant mAbs and the human endogenous IgG Fc region have biantennary complex oligosaccharides with either zero, one, or two terminal galactose moieties, which are the three major glycoforms (G0, G1, or G2) [18,34,43]. The impact of terminal galactose residue on IgG biological functions has been investigated in many studies. Whereas the terminal Gal residue content has shown to play an important role in CDC activity of IgG, the ADCC activity does not seem to be affected by galactosylation of an IgG mAb. Hodoniczky and colleagues have remodeled the Fc *N*-glycans of recombinant therapeutic monoclonal antibody products, Rituxan and Herceptin, in vitro, yielding degalactosylated mAb and other products varying in content of GlcNAc [44]. By degalactosylation of Rituxan and generating mAbs with various Gal content, they have demonstrated that CDC activities and antibody binding to C1q increase as Gal content increases [8,44] (Figure 6). Lower affinity to C1q is due to hydrophobic and hydrophilic interactions between terminal Gal residue and protein, which alter the conformation of the CH2 domain [12]. They confirmed that ADCC activity is not influenced by terminal Gal residue content [44]. Nevertheless, despite some of these contradictory results, galactosylation can induce a positive impact on the binding affinity of the IgG1 to FcγRIIa and FcγRIIIa receptors and ADCC activity [37]. A recent in vitro study also showed that Fc-galactosylation of rituximab enhances CDC activities compared to the degalactosylated glycoform and improvement of C1q binding eventually leads to tumor cell lysis [45]. However, these findings apply to IgG1, but not other subclasses of mAbs. Thus, further detailed, specific investigations of the effects of galactosylation on other IgG subclasses effector functions such as ADCC and CDC are needed.

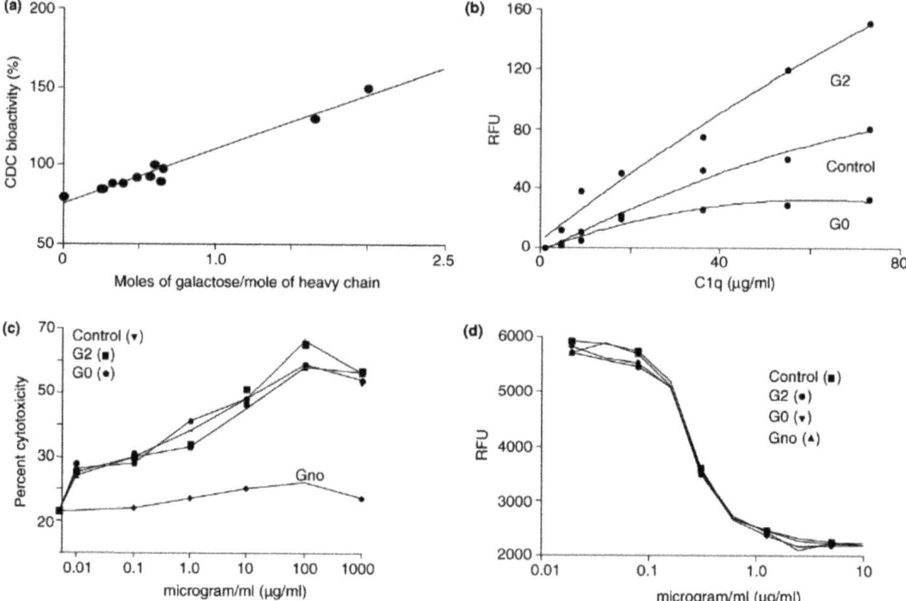

Figure 6. Increase in terminal Gal content increases complement-dependent cytotoxicity (CDC) activity (**a**) and C1q binding (**b**) of rIgGs, but does not affect antibody-dependent cell-mediated cytotoxicity (ADCC) activity (**c**) and antigen binding expressed as relative fluorescence units (RFU) (**d**). G2, G0, and/or Gno (no glycans) glycoforms of Rituxan and/or Herceptin were prepared by in vitro glycosylation methods. These glycoforms, along with control antibody samples (untreated), were subjected to CDC (Rituxan glycoforms), C1q binding (Rituxan glycoforms), ADCC (Herceptin glycoforms), and antigen binding to HER2-ECD (Herceptin glycoforms). Rituxan is a chimeric antibody against CD20 and elicits CDC activity but shows very little ADCC activity. Herceptin is a humanized antibody against HER2-neu antigen and elicits ADCC activity but no CDC activity. Reproduced from Raju 2008 [8] with permission of the copyright owner.

4.3. Bisecting N-Acetylglucosamine

Approximately 10% of human serum endogenous IgGs glycoforms have bisecting GlcNAc residues. Recombinant antibodies generated in CHO cells do not contain bisecting GlcNAc because of the lack of active *N*-acetylglucosaminyltransferase-III (GnT-III) needed for synthesis of bisecting GlcNAc containing *N*-glycans [8,32,46] (Figure 7). Addition of a bisecting GlcNAc has been reported to enhance the binding affinity to FcγRIIIa, which causes 10–30-fold higher ADCC activities [47,48]. Although Hodoniczky et al. [44] approved that bisecting GlcNAc enhances ADCC activity by approximately 10-fold independently of the lack of core fucosylation of rituximab remodeled in vitro, a study done by Shinkawa et al. [48] has debated these findings. Since loss of core fucosylation is always associated with in vivo addition of bisecting GlcNAc, Shinkawa and colleagues proposed that the presence of bisecting GlcNAc may not be the main cause of an ADCC activity increase. As such, Shinkawa's studies demonstrated that the removal of core fucose rather than bisecting GlcNAc has the biggest impact on ADCC activity of therapeutic antibodies [48]. Similar to Shinkawa's results, Ferrara et al. [49] have reported that antibodies enriched in bisected oligosaccharides have increased ADCC.

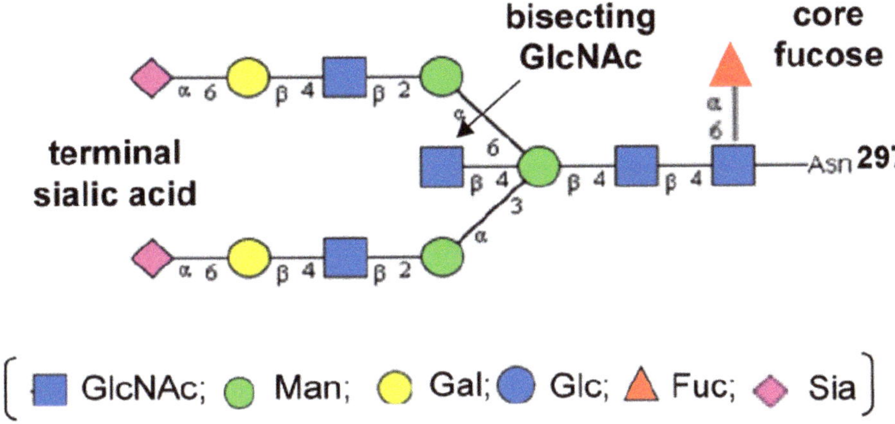

Figure 7. The structure of a full-length biantennary *N*-glycan attached to the Asn-297 in the Fc domain and containing bisecting GlcNAc residue. Reproduced from Huang et al., 2012 [50] with permission of the copyright owner.

4.4. Fucose

The core fucose residues are added to the core GlcNAc residue which is linked to the protein via α-1,6 linkage. This complex glycoform is dominantly found in both serum IgG (>80%) and recombinant IgG's produced in CHO cells (>90%) [32]. Although absence of core fucose residues in Fc glycans has been shown to dramatically improve antibody binding to FcRIIIa and ADCC activity, many studies have demonstrated that the fucosylation level has slight consequences on binding of antibodies to FcγR1, FcγRII, and C1q [48,51,52]. Shields and coworkers used afucosylated anti-HER2 to demonstrate the significant role played by the absence of core fucose in the enhancement of ADCC activity of IgG [51]. About 100-fold greater ADCC exhibited by afucosylated anti-HER2 compared to fucosylated recombinant IgG has been reported in this study. Furthermore, it has been observed that the binding to FcγRI, C1q, or FcRn was not altered [51]. Using marketed nonfucosylated anti-CD20 IgG1 rituximab, Lida et al. confirmed that nonfucosylated IgG1 mediates very high ADCC at low doses in humans, which enhances the therapeutic potential of the modified mAb [53]. Another study has shown that higher binding affinity of afucosylated IgG to Fcγ RIIIa apply for all IgG subclasses [54]. Inclusively, afucosylation of mAb leads the greatest influence on ADCC enhancement, which mediates the efficacy of potential therapeutic recombinant antibodies. Therefore, many recombinant IgGs modified via glycoengineering strategies to generate low-fucose antibodies are currently under investigation in human clinical trials to improve the clinical efficacy of these therapeutics. A classic example of producing low-fucose content antibodies is the anti-CD20 antibody obinutuzumab, which was approved by the US Food and Drug Administration (FDA) in 2013 for the treatment of non-Hodgkin's lymphoma and chronic lymphocytic leukemia. Obinutuzumab showed significantly increased ADCC activity compared with the prototype antibody (rituximab). The same engineered cell line was used to produce a nonfucosylated anti-CD20 antibody (mogamulizumab) that showed a 100-fold increase in ADCC activity compared with the nonglycoengineered rituximab. As such, mogamulizumab was approved for the treatment of adult T cell leukemia/lymphoma in Japan.

4.5. High Mannose

Typically, high mannose content in Fc glycan of IgG varies from five to nine mannose molecules linked to the core GlcNAc. Although about 0.1% of human serum endogenous IgG's contain Fc glycan with high mannose (mostly Man5GlcNAc2 structure), high mannose glycoforms content varies with cell lines and can represent up to 10% of recombinant IgG [20]. Different studies exploring high mannose type Fc glycans have shown that mannose content can impact antibody effector functions. Zhou and his team have demonstrated that the presence of high mannose structures result in enhanced ADCC activity and increased IgG binding affinity for the FcγRIIIa receptor [55]. However, it is unclear whether enhancement in ADCC activity is due to the absence of fucose, since high mannose type glycans possess no core fucose residues [20]. It has also been shown that the IgG with high mannose structures have a negative impact on CDC activity because of lower binding affinity for C1q [55]. Consistently, Kanda et al. have reported similar results that high mannose structures can lead to reduced C1q binding and complement activation [29]. Nevertheless, it was shown that mAbs with high mannose glycan exhibit higher ADCC in the same study. In conclusion, high mannose type Fc glycans have a positive effect on ADCC activity but a negative impact on CDC activity of IgG molecules.

5. Glycoengineering

Since different glycoforms have positive or negative effects on antibody effector functions, it is necessary to develop Fc glycoengineering strategies to facilitate the generation of therapeutic mAbs with consistent and homogenous glycoforms to improve their therapeutic efficacy. Although much progress in cell glycoengineering has already been achieved and important improvements on glycan quality have been accomplished, it is still very challenging to produce IgGs with highly homogenous glycoforms in host cells. The current Fc glycoengineering strategies include host cell glycoengineering and in vitro chemoenzymatic glycosylation remodeling.

5.1. Cell Glycoengineering

Antibody glycosylation is the result of a multistep process. Host cell-based glycoengineering alters glycoforms by genetically modifying important mediators in the glycan biosynthetic pathways to enhance production of desired glycoforms [56]. This technology has been used recently to generate mAbs with optimized quality and efficacy, and focusing on Fc defucosylation which produces a significant increase in ADCC activity results due to the absence of core fucose [57]. Various approaches have been used to modify host cells in order to enhance the desired or limit the unwanted glycoforms. One approach selects host cell type, environmental factors, and cell culture conditions. Host cells that have low FUT8 activity, such as rat hybridoma cell line YB2/0, allow production of recombinant glycoproteins with low core fucose [58,59]. Recombinant mAbs derived from CHO cells exhibit low sialic acid levels because of the absence of α-2,6-sialyltransferase in these cells [60]. Therefore, this cell type is an attractive alternative for the production of mAbs with low sialic acid content. Moreover, cell culture conditions can be modified to favor antibody glycoforms homogeneity. Crowell et al. have reported that feeding the culture with uridine, manganese chloride, and galactose could result in higher CDC activity of mAb due to increased terminal galactose [61]. Another study used 2-fluorofucose, a fucose analogue, to inhibit fucosylation in vitro and produce fucose-deficient antibodies [62].

Another approach in host cells glycoengeneering uses inhibitors of the enzymes that synthesize *N*-linked oligosaccharide chains to alter host biosynthesis pathways. Enzyme inhibitors prevent the addition of outer arm sugar residues including fucose [63]. For example, the addition of ER α-mannosidase inhibitors, deoxymannojirimycin and kifunensine, results in the generation of high mannose (Man9GlcNAc2) glycoform. Another example is that ER glucosidases I and II inhibitors include deoxynojirimycin and castanospermine which arrest mAb in Glc3Man9GlcNAc2 glycoform [63].

A third approach is genetic modulation of the host glycan biosynthesis pathway. This strategy can be performed by upregulating or downregulating substrate expression. Sullivan's group succeeded in the generation of defucosylated antibodies by silencing the *GMD* gene responsible for the expression of GDP fucose, the fucose donor [64]. Furthermore, gene editing techniques, such as ZFNs, TALENs, and CRISPR-Cas9, have been widely used to modify *N*-glycosylation pathways. Chan et al. used these techniques to inactivate the GDP-fucose transporter (SLC35C1) in Chinese hamster ovary (CHO) cells. They concluded that inactivating the *Slc35c1* gene results in production of fucose-free antibodies in CHO cells [65]. Alternatively, small interfering RNis (siRNAs) have been used to knock out multiple genes involved in fucosylation. Finally, inactivation of FUT8 and GDP-mannose 4,6-dehydratase *(GMD)* in CHO cells has led to the production of completely afucosylated IgG with enhanced ADCC [66]. For example, to improve ADCC, a significant improvement through cell-based glycoengineering has been previously reported with the first approved mAbs mogamulizumab and obinutuzumab. Mogamulizumab (POTELIGEO®, KW0761) is a humanized mAb which uses a FUT8 knockout CHO cell line to produce mAbs with nonfucosylated glycan mixtures [66]. Obinutuzumab (Gazyva™, GA-101) is derived from Roche GlycoMAb® technology which overexpresses GnTIII [46,47]. Once the GnT-III adds a bisecting GlcNAc to an oligosaccharide, the core-fucosylation is inhibited. Both technologies produce therapeutic mAbs with enhanced ADCC activity.

5.2. Chemoenzymatic Glycoengineering

Although much successful work in cell glycoengineering has been done to generate therapeutic mAbs with specific glycoforms, it is still very difficult to produce optimized IgGs with homogeneous glycoforms. To accomplish this, chemoenzymatic glycosylation of IgG antibodies provides a new avenue to remodel Fc *N*-glycan from a heterogeneous *N*-glycosylation pattern to a homogeneous one. The Protocol of chemoenzymatic synthesis includes deglycosylation of IgG antibodies using ENG'ase (endo-β-*N*-acetylglucosaminidase) leaving the innermost GlcNAc with or without core fucose at the *N*-glycosylation site. After preparation of glycan oxazolines as donor substrates, a transglycosylation step is used with ENGase-based glycosynthase [66–68] (Figure 8A), and then prepared the glycoengineered mAbs with homogenous *N*-glycans (M3, G0, G2, and A2) via enzymatic reaction (Figure 8B).

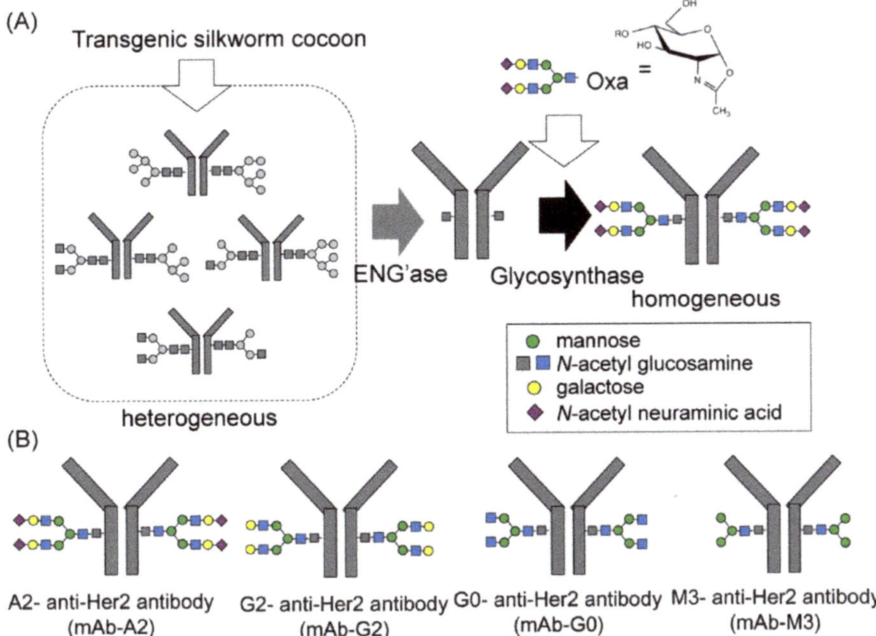

Figure 8. (**A**) Schematic representation of chemoenzymatic synthesis using ENG'ase and glycosynthase. (**B**) Diagram of the homogeneous glycosylated mAb with M3 (mAb-M3), G0 (mAb-G0), G2 (mAb-G2), and A2 (mAb-A2). Reproduced from Kurogochi et al., 2015 [68] with permission of the copyright owner.

There are various ENGases mutants (EndoS D233Q, EndoA N171A, EndoA E173Q, EndoMN175A, and EndoM N175Q) that exhibit transglycosylation activity, which have been engineered to have different substrate specificities and limitations [50,69]. As an example, Huang and coworkers [50] generated two glycosynthase mutants (EndoS-D233A and D233Q) to transform rituximab from mixtures of G0F, G1F, and G2F glycoforms to well-defined homogeneous glycoforms. Using EndoS glycosynthase mutants permitted the production of a fully sialylated (S2G2F) glycoform that shows enhanced anti-inflammatory activity of IVIG's Fc glycans, and a nonfucosylated G2 glycoform that favors increased FcγIIIa receptor-bindings and ADCC activity of mAbs [50] (Figure 9).

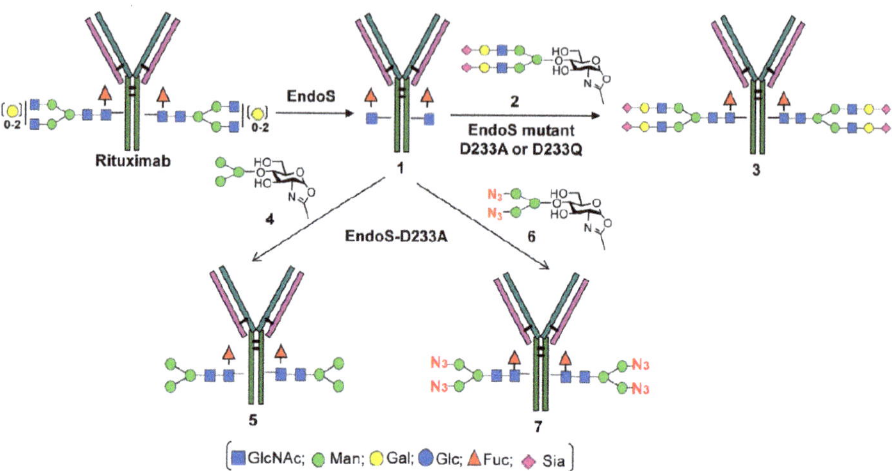

Figure 9. Chemoenzymatic remodeling of rituximab to prepare homogeneous and selectively modified glycoforms. Reproduced from Huang et al., 2012 [50] with permission of the copyright owner.

While many investigations have demonstrated that Endo-S is limited to action on the complex-type, a more recent study described Endo-S2 glycosynthases (D184M and D184Q) that have relaxed substrate specificity and act on transferring three major types (complex, high-mannose, and hybrid type) of *N*-glycans [70]. Collectively, chemoenzymatic glycoengineering technology may be used to develop therapeutic monoclonal antibodies that have homogenous glycoforms, which may circumvent all current efficacy and function quality issues.

5.3. Glycoengineering for Site-Specific Antibody-Drug Conjugation

Antibody-drug conjugates or ADCs are emerging as powerful reagents for the selective delivery of highly toxic drugs to target cells. These relatively novel agents combine the ability of mAbs to bind antigen positive tumor cells with the highly potent killing activity of a cytotoxic drug. In one of the several approaches to obtain structurally-defined, homogeneous antibody–drug conjugates, the Fc glycans of the antibody is engineered for site-specific conjugation [71]. As discussed above, IgGs carry a highly conserved *N*-glycan at the Asn-297 of the Fc domain. Several terminal residues of glycoproteins, including fucose, galactose, and sialic acids that contain vicinal cis diols, can be oxidized selectively with mild periodate (NaIO4) treatment to generate aldehyde groups, which can be further functionalized with other groups, including hydrazides and aminooxy groups for chemoselective conjugation. However, antibody glycosylation is highly heterogeneous, and contains a mixture of galactose and core fucose. As a result, direct oxidation recombinant antibodies usually led to heterogeneous mixtures of the conjugates. For example, to have better control of the homogeneity of ADCs, researchers developed a CHO cell line that could control Fc *N*-glycosylation at the G0F glycoform, where fucose could be selectively oxidized. Thus, treatment of the G0F antibody with mild NaIO4 selectively oxidized the fucose moiety to provide an aldehyde derivative. In contrast to core fucose, oxidation of sialic acid can take place under relatively different conditions because their cis diols are less hindered and more susceptible to periodate oxidation. The advantage of this site-selective modification at the conserved *N*-glycan does not change the IgG structure and thus usually will not affect the antibody's inherent affinity for its antigen. A thorough overview on recent approaches behind glycoengineering of antibodies for site-specific antibody–drug conjugation is described in reference [71].

Today, there are 4 antibody-drug conjugates approved by the US FDA, including Genentech/Roche's Kadcyla® (HER2-specific trastuzumab-drug conjugate) used for the treatment of metastatic breast cancer, Seattle Genetics's Adcetris® (CD30-specific brentuximab-drug conjugate) used for treatment of relapsed Hodgkin's lymphoma, Pfizer/Wyeth's Besponsa™ (CD22-specific inotuzumab-drug conjugate) used for relapsed or refractory B cell precursor acute lymphoblastic leukemia, and more recently, Wyeth Pharmaceuticals' Mylotarg™ (CD33-specific gemtuzumab-ozogamicin conjugate) used for the therapy of acute myelogenous leukemia.

6. Conclusions

In summary, therapeutic mAbs are large, complex, and heterogeneous glycoproteins. They are typically glycosylated at amino acid position 297 in the Fc region. The *N*-glycosylation is crucial for antibody structure and effector functions. The presence or absence of different terminal sugars of Fc glycans can have a significant impact on the PK, PD, and immunogenicity of mAbs (Table 1). Although several studies investigated the correlation between PD and *N*-glycosylation, the results were often contradictory. Whereas high mannose content was shown to significantly impact PK by decreasing antibody half-life, the impact of other glycans on PK is still not fully understood [72]. Collectively, glycosylation is not essential for IgGs' long half-life, and FcRn is the main factor that maintains IgGs' circulation time. Furthermore, therapeutic IgGs derived from non-human cells can be immunogenic as they may express terminal sugar residues that are not naturally found in human serum IgGs, such as sialic acid NGNA. This immunogenicity can decrease the drug efficacy and cause hypersensitivity reactions. For the effects of glycoform patterns on PD and IgG effector functions, the presence of core fucose can interfere with FcγRIIIa binding and ADCC activity of therapeutic antibody. Therefore, the removal of fucose should be considered to enhance ADCC activity of monoclonal antibody drugs. On the other hand, galactosylation can improve the efficacy and quality of mAbs by increasing antibody binding to C1q and CDC of mAbs. Although progress has occurred, there is still much important work to address the unsolved underlying mechanisms that regulate the relationship between changes in Fc-glycan structures and the efficacy and quality of therapeutic monoclonal antibody functions. Due to the critical role of glycosylation and the great impact of different glycans on therapeutic monoclonal antibodies, the need for developing novel glycoengineering strategies has emerged in the last decade [5]. These strategies offer a new route to produce homogenous IgGs with desired glycoforms in order to enhance efficacy and functionality of therapeutic glycoproteins. Glycoengineering techniques which include glycoengineering of cell lines and chemoenzymatic glycoengineering approaches [50,68] are evolving and offer promising novel avenues to develop stable and safer mAbs, which is ultimately linked to lower risk of immunogenicity and higher therapeutic efficacy in humans [5,73–76]. As such, understanding the ways to control the Fc-glycan heterogeneity is essential to the successful clinical development of antibody-based drugs, which can be used to predict their PK/PD during early clinical development and to ensure faster results. This information can appropriately inform manufacturing process development so that these processes are more finely adjusted to deliver the desired Fc glycosylation [77].

Table 1. Summary of potential effects of the most prevalent Fc-glycans on the pharmacokinetics (PK) and pharmacodynamics (PD) of monoclonal antibodies (mAbs). N-Glycolylneuraminic acid form of sialic acid (NGNA).

Fc-Glycans	Potential Effects	References
Fucose	Absence of core fucose enhances: • FcγRIIIa binding • ADCC activity	[48,51–54]
Galactose	Enhances antibody binding to C1q and CDC	[44,45]
Sialic acid	• Anti-inflammatory activity • NGNA reduces FcγRIIIa binding and ADCC activity • NGNA may be immunogenic in human • Removal of sialic acid decreases half-life	[36,39] [35,36] [18–20] [3,25]
High Mannose	• Decreases half-life • Increases FcγRIIIa binding and ADCC activity • Decreases antibody binding to C1q and CDC	[17,25,29] [55] [29,55]
Bisecting GlcNAc	Increases FcγRIIIa binding and ADCC activity	[44,47–49]

In respect to biosimilar development, site-specific glycosylation is also considered crucial in correlating distinct product attributes with observed in vivo effects [13,78]. In this context, an in-depth method for the characterization and analysis of manufactured biosimilar products is required for the production of optimal and consistent biosimilar therapeutic products. A better understanding of the relationship between glycosylation patterns and clinical performance is also of major importance for biosimilar development, and can be used to develop safe and efficacious antibody-based products on the market. In conclusion, antibody glycosylation is necessary to optimize the stability, safety, functionality, and efficacy of therapeutic IgG antibodies. Furthermore, new methods are being applied to generate the next generation of therapeutic mAbs for the treatment of a wide spectrum of human diseases.

To date, antibody-based products are still presenting academia and the biotechnology industry with novel challenges in terms of glycan characterization, stability, and in vivo behavior. Although many studies have been conducted evaluating the various effects of glycosylation on their physicochemical properties and patterns including their importance to biosimilarity [72], our understanding of how glycosylation translates to potential pharmacologic effects and toxicities is still incomplete. Additional investigations are therefore warranted to obtain a clearer pictu re of its importance to antibodies as a vital and important class of drugs.

Author Contributions: S.B. wrote the first draft of the paper as part of her Master Degree Project. All other authors (L.A.K.; A.L.E.; P.H.) contributed substantially to the final text, tables, figure and references to the paper. All authors have read and agreed to the published version of the manuscript.

Funding: This reaerch received no external funding.

Acknowledgments: The authors would like to express their gratitude to Michelle Khawli for her valuable technical editing, language editing, and proofreading. Her willingness to give her time so generously has been very much appreciated.

Conflicts of Interest: All authors declare no conflict of interest.

References

1. Fang, J.; Richardson, J.; Du, Z.; Zhang, Z. Effect of Fc-glycan structure on the conformational stability of IgG revealed by hydrogen/deuterium exchange and limited proteolysis. *Biochemistry* **2016**, *55*, 860–868. [CrossRef] [PubMed]
2. Kohler, G.; Milstein, C. Continuous cultures of fused cells secreting antibody of predefined specificity. *Nature* **1975**, *256*, 495–497. [CrossRef]
3. Liu, L. Antibody glycosylation and its impact on the pharmacokinetics and pharmacodynamics of monoclonal antibodies and Fc-fusion proteins. *J. Pharm. Sci.* **2015**, *104*, 1866–1884. [CrossRef]
4. Bakhtiar, R. Therapeutic recombinant monoclonal antibodies. *J. Chem. Educ.* **2012**, *89*, 1537–1542. [CrossRef]
5. Li, W.; Zhu, Z.; Chen, W.; Feng, Y.; Dimitrov, D.S. Crystallizable fragment glycoengineering for therapeutic antibodies development. *Front. Immunol.* **2017**, *8*, 1554. [CrossRef] [PubMed]
6. Mimura, Y.; Katoh, T.; Saldova, R.; O'Flaherty, R.; Izumi, T.; Mimura-Kimura, Y.; Utsunomiya, T.; Mizukami, Y.; Yamamoto, K.; Matsumoto, T.; et al. Glycosylation engineering of therapeutic IgG antibodies: Challenges for the safety, functionality and efficacy. *Protein Cell* **2018**, *9*, 47–62. [CrossRef]
7. Cymer, F.; Beck, H.; Rohde, A.; Reusch, D. Therapeutic monoclonal antibody N-glycosylation–structure, function and therapeutic potential. *Biologicals* **2018**, *52*, 1–11. [CrossRef] [PubMed]
8. Raju, T.S. Terminal sugars of Fc glycans influence antibody effector functions of IgGs. *Curr. Opin. Immunol.* **2008**, *20*, 471–478. [CrossRef]
9. Butters, T.D. Control in the N-linked glycoprotein biosynthesis pathway. *Chem. Biol.* **2002**, *9*, 1266–1268. [CrossRef]
10. Higel, F.; Seidl, A.; Sörgel, F.; Friess, W. N-glycosylation heterogeneity and the influence on structure, function and pharmacokinetics of monoclonal antibodies and Fc fusion proteins. *Eur. J. Pharm. Biopharm.* **2016**, *100*, 94–100. [CrossRef]
11. Borrok, M.J.; Jung, S.T.; Kang, T.H.; Monzingo, A.F.; Georgiou, G. Revisiting the role of glycosylation in the structure of human IgG Fc. *ACS Chem. Biol.* **2012**, *7*, 1596–1602. [CrossRef]
12. Krapp, S.; Mimura, Y.; Jefferis, R.; Huber, R.; Sondermann, P. Structural analysis of human IgG-Fc glycoforms reveals a correlation between glycosylation and structural integrity. *J. Mol. Biol.* **2003**, *325*, 979–989. [CrossRef]
13. Barbosa, M.D. Immunogenicity of biotherapeutics in the context of developing biosimilars and biobetters. *Drug Discov. Today* **2011**, *16*, 345–353. [CrossRef] [PubMed]
14. Beck, A.; Wagner-Rousset, E.; Bussat, M.-C.; Lokteff, M.; Klinguer-Hamour, C.; Haeuw, J.-F.; Goetsch, L.; Wurch, T.; Dorsselaer, A.; Corvaia, N. Trends in glycosylation, glycoanalysis and glycoengineering of therapeutic antibodies and Fc-fusion proteins. *Curr. Pharm. Biotechnol.* **2008**, *9*, 482–501. [CrossRef] [PubMed]
15. Durocher, Y.; Butler, M. Expression systems for therapeutic glycoprotein production. *Curr. Opin. Biotechnol.* **2009**, *20*, 700–707. [CrossRef]
16. Lam, J.S.; Mansour, M.K.; Specht, C.A.; Levitz, S.M. A model vaccine exploiting fungal mannosylation to increase antigen immunogenicity. *J. Immunol.* **2005**, *175*, 7496–7503. [CrossRef] [PubMed]
17. Goetze, A.M.; Liu, Y.D.; Zhang, Z.; Shah, B.; Lee, E.; Bondarenko, P.V.; Flynn, G.C. High-mannose glycans on the Fc region of therapeutic IgG antibodies increase serum clearance in humans. *Glycobiology* **2011**, *21*, 949–959. [CrossRef] [PubMed]
18. Biburger, M.; Lux, A.; Nimmerjahn, F. How immunoglobulin G antibodies kill target cells. *Adv. Immunol.* **2014**, *124*, 67–94. [PubMed]
19. Tangvoranuntakul, P.; Gagneux, P.; Diaz, S.; Bardor, M.; Varki, N.; Varki, A.; Muchmore, E. Human uptake and incorporation of an immunogenic nonhuman dietary sialic acid. *Proc. Natl. Acad. Sci. USA* **2003**, *100*, 12045–12050. [CrossRef]
20. Qian, J.; Liu, T.; Yang, L.; Daus, A.; Crowley, R.; Zhou, Q. Structural characterization of N-linked oligosaccharides on monoclonal antibody cetuximab by the combination of orthogonal matrix-assisted laser desorption/ionization hybrid quadrupole–quadrupole time-of-flight tandem mass spectrometry and sequential enzymatic digestion. *Anal. Biochem.* **2007**, *364*, 8–18.
21. Liu, L. Pharmacokinetics of monoclonal antibodies and Fc-fusion proteins. *Protein Cell* **2018**, *9*, 15–32. [CrossRef] [PubMed]

22. Roopenian, D.C.; Akilesh, S. FcRn: The neonatal Fc receptor comes of age. *Nat. Rev. Immunol.* **2007**, *7*, 715–725. [CrossRef] [PubMed]
23. Liu, L.; Stadheim, A.; Hamuro, L.; Pittman, T.; Wang, W.; Zha, D.; Hochman, J.; Prueksaritanont, T. Pharmacokinetics of IgG1 monoclonal antibodies produced in humanized Pichia pastoris with specific glycoforms: A comparative study with CHO produced materials. *Biologicals* **2011**, *39*, 205–210. [CrossRef] [PubMed]
24. Chen, X.; Liu, Y.D.; Flynn, G.C. The effect of Fc glycan forms on human IgG2 antibody clearance in humans. *Glycobiology* **2008**, *19*, 240–249. [CrossRef] [PubMed]
25. Yu, M.; Brown, D.; Reed, C.; Chung, S.; Lutman, J.; Stefanich, E.; Wong, A.; Stephan, J.-P.; Bayer, R. Production, characterization and pharmacokinetic properties of antibodies with N-linked Mannose-5 glycans. *MABS* **2012**, *4*, 475–487. [CrossRef]
26. Clarke, S.; Gebbie, C.; Sweeney, C.; Olszewksi, N.; Smith, J. A phase I, pharmacokinetic (PK) and preliminary efficacy assessment of ALD518, a humanized anti-IL-6 antibody, in patients with advanced cancer (Abstract). *J. Clin. Oncol.* **2009**, *27*, 3025.
27. Abuqayyas, L.; Zhang, X.; Balthasar, J.P. Application of knockout mouse models to investigate the influence of FcγR on the pharmacokinetics and anti-platelet effects of MWReg30, a monoclonal anti-GPIIb antibody. *Int. J. Pharm.* **2013**, *444*, 185–192. [CrossRef]
28. Batra, J.; Rathore, A.S. Glycosylation of monoclonal antibody products: Current status and future. *Biotechnol. Prog.* **2016**, *32*, 1091–1102. [CrossRef]
29. Kanda, Y.; Yamada, T.; Mori, K.; Okazaki, A.; Inoue, M.; Kitajima-Miyama, K.; Kuni-Kamochi, R.; Nakano, R.; Yano, K.; Kakita, S.; et al. Comparison of biological activity among nonfucosylated therapeutic IgG1 antibodies with three different N-linked Fc oligosaccharides: The high mannose, hybrid, and complex types. *Glycobiology* **2006**, *17*, 104–118. [CrossRef]
30. Millward, T.A.; Heitzmann, M.; Bill, K.; Längle, U.; Schumacher, P.; Forrer, K. Effect of constant and variable domain glycosylation on pharmacokinetics of therapeutic antibodies in mice. *Biologicals* **2008**, *36*, 41–47. [CrossRef]
31. Sazinsky, S.L.; Ott, R.G.; Silver, N.W.; Tidor, B.; Ravetch, J.V.; Wittrup, K.D. Aglycosylated immunoglobulin G1 variants productively engage activating Fc receptors. *Proc. Natl. Acad. Sci. USA* **2008**, *105*, 20167–20172. [CrossRef] [PubMed]
32. Raju, T.; Briggs, J.B.; Borge, S.M.; Jones, A.J.S. Species-specific variation in glycosylation of IgG: Evidence for the species-specific sialylation and branch-specific galactosylation and importance for engineering recombinant glycoprotein therapeutics. *Glycobiology* **2000**, *10*, 477–486. [CrossRef] [PubMed]
33. Ambrogelly, A.; Gozo, S.; Katiyar, A.; Dellatore, S.; Kune, Y.; Bhat, R.; Sun, J.; Li, N.; Wang, D.; Nowak, C.; et al. Analytical comparability study of recombinant monoclonal antibody therapeutics. *MABS* **2018**, *10*, 513–538. [CrossRef] [PubMed]
34. Flynn, G.C.; Chen, X.; Liu, Y.D.; Shah, B.; Zhang, Z. Naturally occurring glycan forms of human immunoglobulins G1 and G2. *Mol. Immunol.* **2010**, *47*, 2074–2082. [CrossRef]
35. Scallon, B.J.; Tam, S.H.; McCarthy, S.G.; Cai, A.N.; Raju, T.S. Higher levels of sialylated Fc glycans in immunoglobulin G molecules can adversely impact functionality. *Mol. Immunol.* **2007**, *44*, 1524–1534. [CrossRef]
36. Kaneko, Y.; Nimmerjahn, F.; Ravetch, J.V. Anti-Inflammatory activity of immunoglobulin G resulting from Fc sialylation. *Science* **2006**, *313*, 670–673. [CrossRef]
37. Thomann, M.; Schlothauer, T.; Dashivets, T.; Malik, S.; Avenal, C.; Bulau, P.; Rüger, P.; Reusch, D. In vitro glycoengineering of IgG1 and its effect on Fc receptor binding and ADCC activity. *PLoS ONE* **2015**, *10*, e0134949. [CrossRef]
38. Quast, I.; Keller, C.W.; Maurer, M.A.; Giddens, J.P.; Tackenberg, B.; Wang, L.-X.; Münz, C.; Nimmerjahn, F.; Dalakas, M.C.; Lünemann, J.D. Sialylation of IgG Fc domain impairs complement-dependent cytotoxicity. *J. Clin. Investig.* **2015**, *125*, 4160–4170. [CrossRef]
39. Nimmerjahn, F.; Ravetch, J.V. Anti-Inflammatory actions of intravenous immunoglobulin. *Ann. Rev. Immunol.* **2008**, *26*, 513–533. [CrossRef]
40. Anthony, R.M.; Ravetch, J.V. A novel role for the IgG Fc glycan: The anti-inflammatory activity of sialylated IgG Fcs. *J. Clin. Immunol.* **2010**, *30*, 9–14. [CrossRef]

41. Anthony, R.M.; Nimmerjahn, F.; Ashline, D.J.; Reinhold, V.N.; Paulson, J.C.; Ravetch, J.V. Recapitulation of IVIG anti-inflammatory activity with a recombinant IgG Fc. *Science* **2008**, *320*, 373–376. [CrossRef]
42. Bondt, A.; Selman, M.H.J.; Deelder, A.M.; Hazes, J.M.; Willemsen, S.P.; Wuhrer, M.; Dolhain, R.J.E.M. Association between galactosylation of immunoglobulin G and improvement of rheumatoid arthritis during pregnancy is independent of sialylation. *J. Proteome Res.* **2013**, *12*, 4522–4531. [CrossRef] [PubMed]
43. Raju, T.S.; Jordan, R.E. Galactosylation variations in marketed therapeutic antibodies. *MABS* **2012**, *4*, 385–391. [CrossRef] [PubMed]
44. Hodoniczky, J.; Zheng, Y.; James, D. Control of recombinant monoclonal antibody effector functions by Fc N-glycan remodeling in vitro. *Biotechnol. Prog.* **2005**, *21*, 1644–1652. [CrossRef] [PubMed]
45. Peschke, B.; Keller, C.W.; Weber, P.; Quast, I.; Lünemann, J.D. Fc-galactosylation of human immunoglobulin gamma isotypes improves C1q binding and enhances complement-dependent cytotoxicity. *Front. Immunol.* **2017**, *8*, 646. [CrossRef] [PubMed]
46. Patnaik, S.K.; Stanley, P. Lectin-resistant CHO glycosylation mutants. *Meth. Enzymol.* **2006**, *416*, 159–182.
47. Davies, J.; Jiang, L.; Pan, L.-Z.; Labarre, M.J.; Anderson, D.; Reff, M. Expression of GnTIII in a recombinant anti-CD20 CHO production cell line: Expression of antibodies with altered glycoforms leads to an increase in ADCC through higher affinity for FCγRIII. *Biotechnol. Bioeng.* **2001**, *74*, 288–294. [CrossRef]
48. Shinkawa, T.; Nakamura, K.; Yamane, N.; Shoji-Hosaka, E.; Kanda, Y.; Sakurada, M.; Uchida, K.; Anazawa, H.; Satoh, M.; Yamasaki, M.; et al. The absence of fucose but not the presence of galactose or bisecting N-acetylglucosamine of human IgG1 complex-type oligosaccharides shows the critical role of enhancing antibody-dependent cellular cytotoxicity. *J. Biol. Chem.* **2003**, *278*, 3466–3473. [CrossRef]
49. Ferrara, C.; Brünker, P.; Suter, T.; Moser, S.; Püntener, U.; Umaña, P. Modulation of therapeutic antibody effector functions by glycosylation engineering: Influence of Golgi enzyme localization domain and co-expression of heterologous β1, 4-N-acetylglucosaminyltransferase III and Golgi α-mannosidase II. *Biotechnol. Bioeng.* **2006**, *93*, 851–861. [CrossRef]
50. Huang, W.; Giddens, J.; Fan, S.-Q.; Toonstra, C.; Wang, L.-X. Chemoenzymatic glycoengineering of intact IgG antibodies for gain of functions. *J. Am. Chem. Soc.* **2012**, *134*, 12308–12318. [CrossRef]
51. Shields, R.L.; Lai, J.; Keck, R.; Oconnell, L.Y.; Hong, K.; Meng, Y.G.; Weikert, S.H.A.; Presta, L.G. Lack of fucose on human IgG1 N-linked oligosaccharide improves binding to human FcγRIII and antibody-dependent cellular toxicity. *J. Biol. Chem.* **2002**, *277*, 26733–26740. [CrossRef] [PubMed]
52. Niwa, R.; Hatanaka, S.; Shoji-Hosaka, E.; Sakurada, M.; Kobayashi, Y.; Uehara, A.; Yokoi, H.; Nakamura, K.; Shitara, K. Enhancement of the antibody-dependent cellular. *Clin. Cancer Res.* **2004**, *10*, 6248–6255. [CrossRef] [PubMed]
53. Iida, S.; Misaka, H.; Inoue, M.; Shibata, M.; Nakano, R.; Yamane-Ohnuki, N.; Wakitani, M.; Yano, K.; Shitara, K.; Satoh, M. Nonfucosylated therapeutic IgG1 antibody can evade the inhibitory effect of serum immunoglobulin G on antibody-dependent cellular cytotoxicity through its high binding to Fc RIIIa. *Clin. Cancer Res.* **2006**, *12*, 2879–2887. [CrossRef]
54. Niwa, R.; Natsume, A.; Uehara, A.; Wakitani, M.; Iida, S.; Uchida, K.; Satoh, M.; Shitara, K. IgG subclass-independent improvement of antibody-dependent cellular cytotoxicity by fucose removal from Asn297-linked oligosaccharides. *J. Immunol. Meth.* **2005**, *306*, 151–160. [CrossRef] [PubMed]
55. Zhou, Q.; Shankara, S.; Roy, A.; Qiu, H.; Estes, S.; Mcvie-Wylie, A.; Culm-Merdek, K.; Park, A.; Pan, C.; Edmunds, T. Development of a simple and rapid method for producing non-fucosylated oligomannose containing antibodies with increased effector function. *Biotechnol. Bioeng.* **2007**, *99*, 652–665. [CrossRef]
56. Butler, M.; Spearman, M. The choice of mammalian cell host and possibilities for glycosylation engineering. *Curr. Opin. Biotechnol.* **2014**, *30*, 107–112. [CrossRef]
57. Suzuki, E.; Niwa, R.; Saji, S.; Muta, M.; Hirose, M.; Lida, S.; Shiotsu, Y.; Satoh, M.; Shitara, K.; Kondo, M.; et al. A nonfucosylated anti-HER2 antibody augments antibody-dependent cellular cytotoxicity in breast cancer patients. *Clin. Cancer Res.* **2007**, *13*, 1875–1882. [CrossRef]
58. Yamane-Ohnuki, N.; Satoh, M. Production of therapeutic antibodies with controlled fucosylation. *MABS* **2009**, *1*, 230–236. [CrossRef]
59. Urbain, R.; Teillaud, J.L.; Prost, J.F. EMABling antibodies: From feto-maternal allo-immunisation prophylaxis to chronic lymphocytic leukaemia therapy. *Med. Sci.* **2009**, *25*, 1141–1144.
60. Dicker, M.; Strasser, R. Using glyco-engineering to produce therapeutic proteins. *Expert Opin. Biol. Ther.* **2015**, *15*, 1501–1516. [CrossRef]

61. Crowell, C.K.; Grampp, G.E.; Rogers, G.N.; Miller, J.; Scheinman, R.I. Amino acid and manganese supplementation modulates the glycosylation state of erythropoietin in a CHO culture system. *Biotechnol. Bioeng.* **2006**, *96*, 538–549. [CrossRef]
62. Okeley, N.M.; Alley, S.C.; Anderson, M.E.; Boursalian, T.E.; Burke, P.J.; Emmerton, K.M.; Jeffrey, S.C.; Klussman, K.; Law, C.-L.; Sussman, D.; et al. Development of orally active inhibitors of protein and cellular fucosylation. *Proc. Natl. Acad. Sci. USA* **2013**, *110*, 5404–5409. [CrossRef] [PubMed]
63. Powell, L.D. Inhibition of N-Linked Glycosylation. *Curr. Protoc. Mol. Biol.* **1995**, *32*, 17101–17109. [CrossRef]
64. Sullivan, F.X.; Kumar, R.; Kriz, R.; Stahl, M.; Xu, G.-Y.; Rouse, J.; Chang, X.-J.; Boodhoo, A.; Potvin, B.; Cumming, D.A. Molecular cloning of human GDP-mannose 4,6-dehydratase and reconstitution of GDP-fucose biosynthesis in vitro. *J. Biol. Chem.* **1998**, *273*, 8193–8202. [CrossRef]
65. Chan, K.F.; Shahreel, W.; Wan, C.; Teo, G.; Hayati, N.; Tay, S.J.; Tong, W.H.; Yang, Y.; Rudd, P.M.; Zhang, P.; et al. Inactivation of GDP-fucose transporter gene (Slc35c1) in CHO cells by ZFNs, TALENs and CRISPR-Cas9 for production of fucose-free antibodies. *Biotechnol. J.* **2015**, *11*, 399–414. [CrossRef] [PubMed]
66. Imai-Nishiya, H.; Mori, K.; Inoue, M.; Wakitani, M.; Lida, S.; Shitara, K.; Satoh, M. Double knockdown of alpha 1,6-fucosyltransferase (FUT8) and GDP-mannose 4,6-dehydratase (GMD) in antibody-producing cells: A new strategy for generating fully non-fucosylated therapeutic antibodies with enhanced ADCC. *BMC Biotechnol.* **2007**, *7*, 84. [CrossRef] [PubMed]
67. Giddens, J.P.; Wang, L.-X. Chemoenzymatic Glyco-engineering of monoclonal antibodies. *Methods Mol. Biol.* **2015**, *1321*, 375–387. [PubMed]
68. Kurogochi, M.; Mori, M.; Osumi, K.; Tojino, M.; Sugawara, S.-I.; Takashima, S.; Hirose, Y.; Tsukimura, W.; Mizuno, M.; Amano, J.; et al. Glycoengineered monoclonal antibodies with homogeneous glycan (M3, G0, G2, and A2) using a chemoenzymatic approach have different affinities for FcγRIIIa and variable antibody-dependent cellular cytotoxicity activities. *PLoS ONE* **2015**, *10*, e0132848. [CrossRef]
69. Umekawa, M.; Huang, W.; Li, B.; Fujita, K.; Ashida, H.; Wang, L.-X.; Yamamoto, K. Mutants of mucor hiemalis endo-β-N-acetylglucosaminidase show enhanced transglycosylation and glycosynthase-like activities. *J. Biol. Chem.* **2007**, *283*, 4469–4479. [CrossRef]
70. Li, T.; Tong, X.; Yang, Q.; Giddens, J.P.; Wang, L.-X. Glycosynthase mutants of endoglycosidase S2 show potent transglycosylation activity and remarkably relaxed substrate specificity for antibody glycosylation remodeling. *J. Biol. Chem.* **2016**, *291*, 16508–16518. [CrossRef]
71. Wang, L.-X.; Tong, X.; Li, C.; Giddens, J.P.; Li, T. Glycoengineering of Antibodies for Modulating Functions. *Annu. Rev. Biochem.* **2019**, *88*, 433–459. [CrossRef]
72. Bumbaca, D.; Boswell, C.A.; Fielder, P.J.; Khawli, L.A. Physiochemical and biochemical factors influencing the pharmacokinetics of antibody therapeutics. *AAPS J.* **2012**, *14*, 554–558. [CrossRef] [PubMed]
73. Jefferis, R. Glycosylation as a strategy to improve antibody based therapeutics. *Nat. Rev. Drug Discov.* **2009**, *8*, 226–234. [CrossRef] [PubMed]
74. Smith, A.; Manoli, H.; Jaw, S.; Frutoz, K.; Epstein, A.L.; Theil, F.P.; Khawli, L.A. Unraveling the effect of immunogenicity response on the PK/PD, efficacy, and safety of biologics. *J. Immunol. Res.* **2016**, *2016*, 9. [CrossRef] [PubMed]
75. Lu, Y.; Khawli, L.A.; Purushothama, S.; Theil, F.P.; Partridge, M. Recent advances in assessing immunogenicity of therapeutic proteins: Impact on biotherapeutic development. *J. Immunol. Res.* **2016**, *2016*, 1–2. [CrossRef] [PubMed]
76. Beck, A.; Liu, H. Macro-and micro-heterogeneity of natural and recombinant IgG antibodies. *Antibodies* **2019**, *8*, 18. [CrossRef] [PubMed]
77. Reusch, D.; Tejada, M.L. Fc glycans of therapeutic antibodies as critical quality attributes. *Glycobiology* **2015**, *25*, 1325–1334. [CrossRef] [PubMed]
78. Duivelshof, B.L.; Jiskoot, W.; Beck, A.; Veuthey, J.-L.; Guillarme, D.; D'Atri, V. Glycosylation of biosimilars: Recent advances in analytical characterization and clinical implications. *Anal. Chim. Acta* **2019**, *1089*, 1–18. [CrossRef]

MDPI
St. Alban-Anlage 66
4052 Basel
Switzerland
Tel. +41 61 683 77 34
Fax +41 61 302 89 18
www.mdpi.com

Antibodies Editorial Office
E-mail: antibodies@mdpi.com
www.mdpi.com/journal/antibodies

www.ingramcontent.com/pod-product-compliance
Lightning Source LLC
LaVergne TN
LVHW070652170726
843515LV00004B/386

* 9 7 8 3 0 3 6 5 2 8 7 3 1 *